A Watershed Year

A Bur Oak Book

A Watershed Year

Anatomy of the Iowa Floods of 2008

edited by Cornelia F. Mutel

UNIVERSITY OF IOWA PRESS ≋ IOWA CITY

University of Iowa Press, Iowa City 52242
Copyright © 2010 by the University of Iowa Press
www.uiowapress.org
Printed in the United States of America
Design by Sara T. Sauers

No part of this book may be reproduced or used in any form
or by any means without permission in writing from the
publisher. All reasonable steps have been taken to contact
copyright holders of material used in this book. The publisher
would be pleased to make suitable arrangements with any
whom it has not been possible to reach.

Supported by the Iowa Flood Center and by grants from IIHR-
Hydroscience & Engineering and the Center for Global and
Regional Environmental Research, the University of Iowa.

The University of Iowa Press is a member of Green Press
Initiative and is committed to preserving natural resources.

Printed on acid-free paper

Library of Congress Cataloging-in-Publication Data

A watershed year: anatomy of the Iowa floods of 2008 /
edited by Cornelia F. Mutel.
          p.   cm.
Includes bibliographical references and index.
ISBN-13: 978-1-58729-854-7 (pbk.)
ISBN-10: 1-58729-854-6 (pbk.)
1. Floods—Iowa—History.   I. Mutel, Cornelia Fleischer.
GB1399.4.I6W37   2010
363.34'9309777—dc22   2009024948

Contents

Preface and Acknowledgments

~~~~~~

IN EARLY JUNE 2008, my husband and I left our home near Iowa City to visit our grandchildren and their parents in Switzerland. The skies were crystalline blue and cloudless, the air fresh. Crossing the Coralville Reservoir on the way to the airport, I commented that the water level was unusually high, but gave the matter no more thought. By the time our plane landed and we had slept off the trip and turned on the computer, the news from home was dominated by flood fears. I was stunned. For the next several weeks, I stared obsessively at the computer screen watching videos of rising waters and their effects, reading flow predictions and stories of human woe. I directed a long-distance evacuation of my university office, which overlooks the Iowa River. And I felt horrible— horrible about not being there to help with the sandbagging and evacuations. And horrible about missing out on a larger-than-life event that was going on in my own backyard. Throughout my life, I have been a student of the natural world, studying native plant communities even as I listened to nature's quieter voices, trying to discern what they said about process and change. Now process and change were occurring at breakneck speed throughout eastern Iowa, and I wasn't there to witness.

We returned to a post-flood world struggling to recover from shock and devastation. I volunteered to help feed hundreds of abandoned pets stashed in horse stalls at a nearby community college. I helped gut a flooded house, carrying to the curb moldy dolls and water-stained pink ballet skirts eerily similar to my granddaughter's. I listened to friends—one who told me how the entire inventory of his small business had been swept down the Cedar River, another who said that her boyfriend was moving in for good because the river had taken his house. These were stories of lives changed forever in an instant. And I—like thousands of other eastern Iowans—wondered how to cope with the mixture of personal emotions and life-reshaping changes that were swirling within and around me.

This book was the result of these observations and emotions. It started out as my attempt to help my community deal with the floods and evolved from there. I planned to produce a document that would clear up confusion, address complexity, and confront misinformation by explaining the science and facts necessary for dealing with future floods and recovering from this one. I intended to edit a book that would be usable by the experts, yet understandable to the lay public and students of natural disasters as well as administrators, land managers, and policy makers. The potential audience would thus be broad, including anyone with an interest in the subject or working with these types of natural disasters. I thought that the book could become a standard of reference for the 2008 floods and for future floods that are sure to come. But I hoped that more than this, it would stimulate discourse and feed into policy decisions by broadening the public's vision of flood processes, causes, costs, and remediation.

Editing this book was far more complex and time-consuming than I originally assumed. My first job was finding and engaging chapter authors—which meant defining book content. The latter continued to evolve as engaged authors revised their chapters and as new authors joined the project. Early on, I decided to concentrate on the scientific and fact-based aspects of the floods but to minimize discussion of their social and policy aspects, important topics but too large to be folded neatly into this science-based project. I also conceived of the book as a relatively rapid response to the flooding. As such, I realized that it would not be comprehensive and that chapters would sometimes be based on preliminary data and research. I thus acknowledged that conclusive and comprehensive works on the floods could not be published for years hence.

I sought authors on the basis of their expertise on a given subject. Many are researchers or land use managers who are regional or national experts in their fields. Some are administrators working onsite to manage the flood response. All had intimate experience with flooding. I guided their writing, encouraging an approachable style and simple explanations of technical matters. The authors and I passed chapters back and forth, often several times, as we worked to perfect both form and content. As we did, we made many attempts to check the accuracy of information throughout, to ensure that similar databases were used from chapter to chapter and to cross-check chapters for consistency. This was more easily said than done: flood data were continuously evolving even as we wrote, and different sources of information sometimes presented different data for the same event. As editor, I also worked to point out major themes and tie together the book's diverse content by writing introductions to the book and to each of the book's four sections, as well as an epilogue. When the drafts of all the contributions were finally completed, content experts reviewed chapters in the area of their expertise. Upon completion of the draft manuscript, active and experienced flood researchers reviewed the entire document. Graphics and tables were collected or created in parallel with the text, receiving the same intensive review.

This book is thus the result of the dedication and effort of dozens of authors, artists, reviewers, and supporters, all of whom won my esteem and eternal gratitude. First and foremost, I express my most sincere thanks to the authors themselves, who met my requests for rewrites time (and time, and time) again. I'm sure that often they inwardly groaned at my approach, yet they unfailingly responded with grace and creativity. Sara Sauers, graphics editor and designer for the book, also was crucial to its production, expertly selecting photos and preparing graphics for publication. I could not have done it without her. Michael Kundert and Radoslaw Goska (both at IIHR-Hydroscience & Engineering) and Casey Kohrt (Iowa Geological and Water Survey) donated or executed illustrations for the book; they too responded with grace to multiple edits. Allen Bradley not only wrote two of the chapters, he also served as my mentor and regular source of hydrologic information. He clarified many issues, answered multiple questions, and reviewed material for me repeatedly, significantly improving the book's content. David Eash, another chapter author, also was extraordinarily gracious in repeatedly explaining technicalities to me. Robert Ettema provided me with ideas and information. Holly Carver, director of the University of Iowa Press, was an unflinching source of encouragement and belief in the project's potential. Linzee McCray's careful eye examined the entire manuscript; her edits improved clarity and content. Other reviewers of both individual chapters and the entire book made useful and much-appreciated comments. The book was executed under funding from and the support of my home institution, IIHR-Hydroscience & Engineering, and the Center for Global and Regional Environmental Research, both at the University of Iowa. To all these persons and institutions, I express my heartfelt thanks.

Despite everyone's efforts to ensure consistency and accuracy, this project was as complex as it was diverse. I accept responsibility for any errors that may have crept into the final product and express my apologies to both readers and the authors for any such items that I failed to catch.

Introduction

In June of 2008, the rivers of eastern Iowa rose above their banks to create floods of epic proportions (fig. I-1). Onlookers noted not only their amazing size—the flow through Cedar Rapids for example reaching 140,000 cubic feet per second (cfs), nearly double the earlier record flood flow in 1961—but also the rapidity of their rise. Each day posed new and previously unimaginable threats. In rural areas, floods ruined farmlands either by causing massive erosion or by depositing layers of sand and rock. They closed transportation routes across Iowa. In Cedar Rapids, the flood displaced thousands of residents and hundreds of businesses and inundated over nine square miles of the downtown area. Iowa City, including the University of Iowa and the adjacent town of Coralville (just northwest of Iowa City), also suffered major flood damage, although the Coralville Reservoir and Dam moderated flows here. While the flood in Iowa City was the most destructive in recorded history, it was not the largest on record. In 1851, a century before the building of the Coralville Dam, the Iowa River flowed at 70,000 cfs through a much younger Iowa City, a discharge nearly twice as large as the 2008 flow of 41,100 cfs, and in 1881 it flowed through town at 51,000 cfs. Indeed, large floods undoubtedly occurred even before the Euroamerican settlement of Iowa and the plowing of the prairie. But these facts did not lessen the shock for modern residents who remembered extreme floods in the same region just 15 years earlier, in 1993. Surely this could not be happening again, and so soon?

Much of the extensive media coverage stressed the social aspects of the

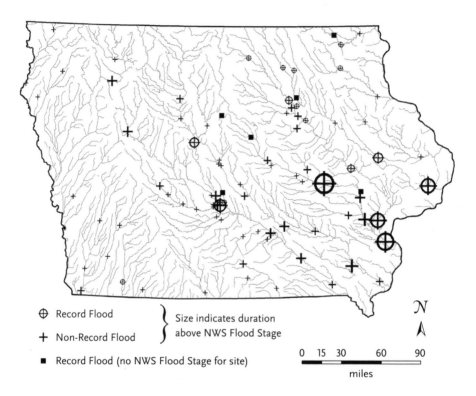

FIGURE I-1 In June 2008, floods of historic proportions raged across the Iowa landscape. As shown in this map, the flooding affected all corners of the state, although floods were largest and longest lasting in eastern Iowa, where many were the largest ever seen ("record floods," shown by circles and black squares). Some of these very large floods were also exceptionally long-lasting (as indicated by larger map symbols), continuing for weeks. Over half (83) of the state's total number of Iowa's U.S. Geological Survey (USGS) streamgages recorded water levels above the National Weather Service (NWS) flood stage. Unprecedented flood levels caused catastrophic damages, and 85 of Iowa's 99 counties were declared federal disaster areas by FEMA. Note that this map shows streamgage sites only; undoubtedly many other locations also flooded but had no stream monitoring devices to record the event.
Illustration by Mohamed Habib and A. Allen Bradley, Jr., based on USGS Iowa Water Science Center (http://ia.water.usgs.gov/), record flooding data; and Monthly Flood Reports, USGS WaterWatch (http://water.usgs.gov/waterwatch/), duration above flood stage.

flooding and the extent of property damage. But other stories were also being told, albeit in a quieter and less publicized manner. Hydrologists were struggling to measure and characterize the floodwaters. Scientists at the University of Iowa were testing new methods of analyzing flood data and forecasting flood flows. And researchers and the public alike were hypothesizing about their cause. Did Iowa's agricultural and urban landscape magnify the water's rapid rises? Could changing climate have played a role? The media told fragments of these and other science-based stories and used technical jargon liberally, if not always correctly. Soon estimates of cost started to emerge, along with statements about how to prevent the next flood. But the public had few tools with which to understand or assess any of these matters.

This book attempts to remedy this situation by providing a solid base of flood-related information. It encourages readers to examine relationships among rivers, floodplains, weather, and modern society, stressing matters of science and fact rather than social or policy issues. The style is intended to present science and technical matters in a manner that is understandable and interesting to most educated lay readers while still being useful to experts. Chapters were written by 30 authors who are experts in their fields, many of whom dealt first-hand with the 2008 floods. The book thus provides a close-up view of flood events and their impacts. It should intrigue Iowans who experienced the 2008 floods, but with its breadth of information, the book also will be useful to a variety of readers concerned with flooding more generally. The goal is not only to educate these readers but also to stimulate discussion and wiser land use, flood-related responses, and policies in coming years.

The book looks in detail at two regions that were intensely affected by the floods, the Iowa and Cedar River watersheds, which stretch from east-central Iowa northward into southern Minnesota (fig. I-2). These watersheds suffered greater damage from the 2008 floods than from any other natural disaster in recorded history and thus provide ample examples and data for bringing extreme flooding to life. Within these watersheds, we focus on Linn and Johnson Counties, sites of two of Iowa's largest cities (Cedar Rapids and Iowa City/Coralville). These particular counties and cities allow comparisons that further elucidate flood processes. The Cedar River is free-flowing through Cedar Rapids, down to its junction with the Iowa River at Columbus Junction. Cedar Rapids has extensive residential, industrial, and business districts on the broad floodplain of the Cedar River. The Iowa River, in contrast, is regulated by the Coralville Dam and Reservoir below Marengo. Iowa City is partially protected by this reservoir, with the city's relatively small floodplain and downtown area lying

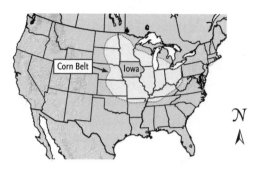

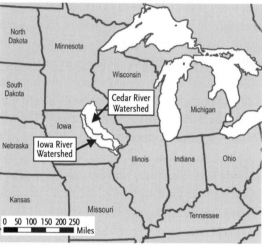

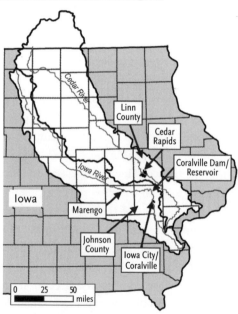

FIGURE I-2 This book focuses on eastern Iowa's Cedar and Iowa River watersheds, which with their signature cities Cedar Rapids and Iowa City/Coralville were the hardest hit by the June 2008 floods. Examining this region in depth enabled the authors to create a detailed picture of a single series of floods, their causation, and their impact. While this book uses these floods and locations as examples, an understanding of this flood event will allow readers to better respond to floods throughout the midwestern Corn Belt and beyond. *Illustration by Michael Kundert.*

about 8 river miles downstream from the dam. Iowa City also is the home of the University of Iowa, which spans both banks of the river. Many university buildings were seriously flooded in 2008, including offices of one of the nation's preeminent hydraulic and hydrology research institutes, IIHR-Hydroscience & Engineering. Because of this institute, research has for many years been focused on water's movement through the Iowa River watershed, and this river will continue to be a hydrologic research focus—another reason for writing about these particular locations. The book's concentration on these particular cities, counties, and watersheds is not meant in any way to minimize the significant flood damage and difficulties faced by other cities or by Iowa's rural lands and smaller towns in 2008.

While the book draws most of its examples and stories from this one particular region, it does so with the intent of explaining flooding throughout a much larger area: the midwestern Corn Belt, which stretches from eastern Nebraska into Ohio (see fig. I-2). The Corn Belt, once the domain of North America's midcontinental tallgrass prairie, is defined by its deep fertile soils, gentle terrain, well-distributed rainfall, and long hot summers, traits that foster crop growth. With these attributes, the Corn Belt has become one of the most intensively farmed and managed agricultural landscapes in the world. While major cities such as Iowa City and Cedar Rapids dot the open expanses, the majority of this primarily rural region (and about two-thirds of Iowa) is now dedicated to annual row crops, primarily corn and soybeans. This farming practice, along with the landscape modifications that transformed prairie into cropland, has robbed the land of much of its original water-holding capacity. Soils lie bare for much of the year, captive to the heavy rains and intense runoff that characterize spring and early summer and led to the 2008 Iowa floods. The Iowa and Cedar River watersheds lie in the heart of the Corn Belt, and the flooding processes, results, and remediation suggestions explained in this book hold true throughout this vast landscape. Section IV and other scattered chapters apply well beyond the Midwest—indeed throughout the nation.

The book's 25 chapters fall naturally into four sections. "Rising Rivers, Spreading Waters" begins with a basic description of flooding in Iowa, told in part by comparing the 1993 and 2008 floods. Following chapters examine flood hydrology (the science of the water cycle), forecasts and forecasting methods, and determinations of flood size and frequency and also tell of community responses to the 2008 floods. "Why Here, Why Now?" searches for possible causes of the floods, considering those that were most frequently cited in 2008: land use, releases from the Coralville Dam on the Iowa River, and climate change.

"Flood Damages, Flood Costs, Flood Benefits" looks at the results of the floods, emphasizing damages to riverside cities and farmland, but also describes potential benefits of floods to natural communities and archaeological sites. The book ends with "Looking Back, Looking Forward," which lays out approaches we can take to decrease the danger and damage of future floods.

Because the words "river" and "flood" are integral to every page of this book, let's consider their meaning, which is not as simple as one might assume. We think of rivers as we normally see them: flows of water confined to well-delineated channels. When a river rises above its banks and spreads over surrounding flatlands, we see it as an unnatural or abnormal state for the river. We may understand rivers better if we view them according to their function—that is, as geologic agents reshaping the land by cutting their way through the landscape en route to a larger river, lake, or the sea. A river serves as a conveyor belt that is drying its watershed (also called drainage basin, the area comprised of all upstream areas feeding into a central channel) by transporting water and sediment to the sea. The mineral particles and accompanying organic matter that make up sediment are components of all rivers. They become major river components in the Midwest, where water cuts through deep soils.

Rivers are marked by constant change; they continually erode and deposit sediment, reshaping and resizing their channels in the process. Rivers work to establish a balance between their flow, the sediment load in the flow, and the size and slope of their channel, achieving an equilibrium that keeps the river in its main channel. But sometimes water flowing into the river exceeds the river channel's carrying capacity, and more room is needed. Rivers carve larger auxiliary channels to carry and partially store this excess water. These auxiliary channels flank the main channel, forming what is called the river's floodplain. The entire floodplain is part of the extended river channel, standing by to carry and store water when it becomes necessary, with sections that are lowest and most easily connected to the river flooding first and most frequently. Floodplains are thus necessary parts of all properly functioning rivers, even though they are not constantly occupied by water.

Seen in this way, the excessive river flow that we define as a flood is not abnormal. It is as natural as the rising of the sun—and no more easily prevented than the earth's circling of the sun. Floods are what rivers do. Floods are one component of the water cycle, a process as ancient and necessary as any of nature's cycles. Wherever people or property are found close to a river, the potential for flood damage exists. Floods become a problem only because we choose to live and place objects of value in a river's extended channel—that is, in its floodplain.

A few words about using this book. While the ideas are meant to flow from one chapter to the next, each chapter has been written to stand independently, with technical terms and acronyms defined the first time they are used in all chapters. Captions have also been designed to allow the figures an independence from the text. Several chapters have color plates that are included in a color insert. These graphics are integral parts of the chapters; readers will do well to refer to the color insert whenever plates are referenced in the text. Figures I-3 and I-4 locate some of the sites that are mentioned in the text; these figures, figure I-2, and plate 1 should be used as references throughout the book.

Occasional footnotes explain technical matters that will interest some readers but are deemed too technical for the chapter text. Because flood data are complex and may differ among different sites, we were careful to cite information sources throughout the chapters and for graphics, either as References Cited (at each chapter's end) or in the graphics credit lines. Much information was gathered on the web; the date of accessing websites was included only when data on those sites are continually being added or changing. We have used the word "flood" (singular) to refer to a rise of water above the river's banks at a single location (the flood at Cedar Rapids) or within a single river drainage (the flood on the Cedar River), with the word "floods" referring to rises in multiple drainages (the floods in Iowa) or at different times in history (past floods at Cedar Rapids). A "flood event" is the result of specific weather conditions (for example, a certain set of storms), which usually affect multiple drainages and thus cause floods.

Note that we edited chapters with an eye to ensuring consistency of facts and terminology but not necessarily to eliminating diverging opinions. Science is a process that thrives on healthy discourse, especially with subjects as massive and complex as floods. Ambiguity is constant and must be accepted as part of the process. Different perspectives leading to apparently conflicting outcomes are not to be discouraged, for this is precisely the way that new knowledge is moved forward. This premise holds equally true for political, policy, and social concerns addressing flooding issues. Healthy discourse and open communication are crucial for developing the best science; arriving at the optimal mix of engineering, policy, and land-use decisions; promoting an informed public; and ultimately creating a future that lessens flood damage and danger.

This book is intended to spark flood-related explorations, not summarize endpoints. My hope is that it will stimulate discourse that will, in turn, spur the creation of new knowledge as well as better understanding. With these tools, we will be able to craft a wiser relationship with our rivers and land, one that in coming years decreases the threat of floods' destructive power, enhances the health of our environment, and is mutually beneficial for both people and the many other species that cohabit our planet.

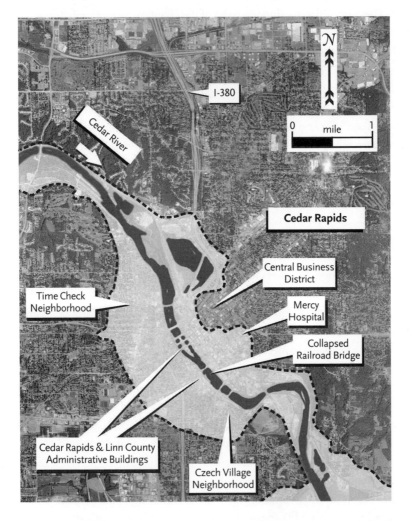

FIGURE I-3 Selected sites in Cedar Rapids mentioned in the text. Flooded areas are shown in light gray; the main river channel and lakes are shown in dark gray. *Illustration by Michael Kundert, drawn from Linn County Emergency Management Agency flood map, available at http://www.linncounty.org. Background photo from ISU GIS Facility, http://ortho.gis.iastate.edu/.*

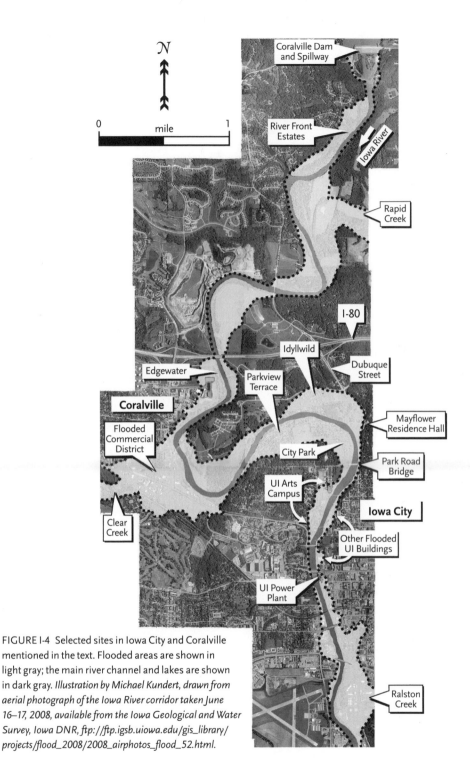

N

0 mile 1

Coralville Dam
and Spillway

River Front
Estates

Iowa River

Rapid
Creek

I-80

Idyllwild

Dubuque
Street

Edgewater

Parkview
Terrace

Coralville

Mayflower
Residence Hall

Flooded
Commercial
District

City Park

Park Road
Bridge

UI Arts
Campus

Iowa City

Clear
Creek

Other Flooded
UI Buildings

UI Power
Plant

Ralston
Creek

FIGURE I-4 Selected sites in Iowa City and Coralville
mentioned in the text. Flooded areas are shown in
light gray; the main river channel and lakes are shown
in dark gray. *Illustration by Michael Kundert, drawn from
aerial photograph of the Iowa River corridor taken June
16–17, 2008, available from the Iowa Geological and Water
Survey, Iowa DNR, ftp://ftp.igsb.uiowa.edu/gis_library/
projects/flood_2008/2008_airphotos_flood_52.html.*

Rising Rivers, Spreading Waters

N ATURE SPEAKS TO US in many ways. We watch, listen, and try to understand. Songbirds returning in the spring tell of homecoming and new life. Whales sing of ocean swells and planetary migrations. Some voices, like those of Arctic mammals swimming toward melting ice floes, speak of a planet where conditions seem to be changing too rapidly for their own good—and for ours. Midwesterners may think that they hear similar suggestions of Earth's processes out of kilter when spring rainstorms thunder over the Midwest, storms that waken memories of previous years when waters rose too fast and spread too far.

This happened in eastern Iowa in June of 2008. Those who listened heard proclamations of nature's awesome power and intensity, of the complexity and magnitude of Earth's processes. But they also smelled the stench of the floods and saw the unexpected responses of other creatures—water birds flourishing in the havoc and hundreds of drowned garter snakes littering mud-coated sidewalks once the waters receded. They heard the emotional and financial tragedies of neighbors who lost homes, businesses, and livelihoods. And they marveled at the mountains of crumbling wallboard, molding wood, ruined clothes and toys, and discolored appliances that appeared along streets as residents gutted their homes.

This section speaks of the 2008 floods in two very different manners. In chapters 3, 4, and 5, Richard A. Fosse, Barbara Eckstein and Rodney Lehnertz, and Linda Langston detail the responses of Iowa City, the University of Iowa, and Linn County to the rising 2008 waters. These chapters, written by community leaders who were on site throughout the flood period, give a vivid firsthand picture of the way Iowans struggled with nature's extremes, incomplete information, and changing flood predictions and came together to meet the threat in a typically Iowa way, with emphasis on concern for others and community integrity.

Chapters 1, 2, 6, and 7 describe the floods scientifically, as physical processes to be explored and explained. This approach relies on collecting facts about the rising waters, in particular their levels and flow rates. To understand these chapters, we need to know something about the federal government's streamgage system, which routinely collects this information. This series of streamgage data collection sites serves as our eyes on the rivers (plate 1).[1] Without their routine measurements, our consideration of river behavior including floods would be largely guesswork.

Streamgages continuously collect information on a river's stage, its water level above a base elevation such as a streambed. These measurements, when combined with periodic measurements of the river's discharge, that is, its flow rate expressed in cubic feet per second or cfs, permit the translation of water level into continuous discharge records. These data are plotted to create hydrographs of a river's changes through time. The 2008 stage hydrographs, graphs of river height, are included in chapter 6, figure 6-1; chapter 1, figure 1-4 is a hydrograph of the Coralville Reservoir elevation during the 2008 and 1993 flood years. Discharge hydrographs, graphs of river flow, are included in chapter 1, figure 1-5, and chapter 2, figures 2-1, 2-2, and 2-3.

The record of river discharges allows hydrologists, those who study water's flow, to characterize a given river site, track a river's changes over time, and compare one river to another and one site on a river to upstream and downstream sites. The data also are linked to computer models to predict future river stage and discharge values, both during normal flows and for floodwaters. Note that river data are available only for sites with streamgages; information about flows at other river sites must be extrapolated using data from nearby streamgages. River discharges can vary greatly from one river site to another and from one river to another (Blanchard 2007).

Streamgage and weather data are cited repeatedly throughout chapters 1, 2, 6, and 7. The section begins with A. Allen Bradley, Jr., summarizing the general traits of Iowa floods (chapter 1). He discusses the season and size of floods on rivers of different magnitudes, explaining, for example, why nearly all Iowa floods occur in spring and summer, and pointing out that spring flooding is more common on larger rivers but summer flooding is more common on smaller rivers. He then proceeds to examine Iowa's biggest floods by comparing those of 1993 and 2008.

In chapter 2, Witold F. Krajewski and Ricardo Mantilla address questions such as "why were the floods so large?" The authors show that the heavy rains only partially explained the flooding. While the weather in 2008 (with its snowy winter, cool moist spring, and heavy early summer rainstorms) was prime for flooding (see figure 2-3, also plate 14), the year's record floods are best explained through a new approach to analyzing flood data that these researchers are investigating, which examines the merging of rainfalls and river flows over time. Heavy rains falling high in the Cedar Rapids watershed ran into rivers that flowed downstream to merge, a few days later, with runoff from heavy rains falling lower in the watershed, creating the massive swell that spilled out to cover Cedar Rapids.

Chapter 6 talks about the tricky job of looking into the future to forecast the peak of a flood that, in a given location, is rising higher than anyone has seen there before. Such forecasts and flood warnings are fraught with difficulty, but they are crucial to communities that are threatened by flood events. A. Allen Bradley, Jr., outlines the process of making flood forecasts and explains how this was done for the 2008 floods. The second half of his chapter chronologically describes the floods' hydrology in a very understandable manner.

Chapter 7, on flood frequencies, defines the oft-used but poorly understood terms "100-year flood" and "500-year flood." These are, David Eash explains, statements about the size of a flood, expressed as the probability that a flood of a certain size will occur in any given year. For example, a 100-year flood has a 1 percent chance of occurring this year, and next year, and the next. Thus multiple 100-year floods can indeed occur in close succession. These probabilities gain validity and continue to be reset as data records become longer. Eash describes this process. The chapter includes graphs of the size of historic floods in Iowa City and Cedar Rapids (figures 7-2 and 7-3), as well as a very helpful table of discharges and flood records (including flood frequency designations) at streamgage stations in the two watersheds under discussion.

The community-oriented chapters of this first section grab our interest and touch our spirits. Chapters presenting technical information on floods provide us with valuable tools for quantifying, forecasting, and understanding floods and their mechanisms. Thus this nuts-and-bolts section provides the underlayment for the rest of the book, as well as for flood readings elsewhere. It summarizes statistics on the 2008 floods and amply displays them through graphs and tables. The section is sure to provide a handy reference to anyone who continues to consider the 2008 floods and a solid factual basis for moving into section II. ≋

Note

1. Streamgage stations consist of small shelters that hold devices to measure and record stage and feed the data back to a central office via satellite links. Data for most are now distributed in near-real time over the Internet. The nation's streamgages are operated by the U.S. Geological Survey, which provides river data to the National Weather Service, U.S. Army Corps of Engineers, and other government agencies. However, most are funded cooperatively through cost-sharing with state or local governments or agencies. Iowa has 154 streamgage stations that have been operating, on average, 50 to 60 years. Thirteen of these streamgages are on the Iowa and Cedar Rivers in Iowa (see plate 1), with additional streamgages in these watersheds being located on smaller tributaries and in Minnesota. Iowa's earliest streamgage was established in 1873 on the Mississippi River at Clinton (Lund 2009).

References Cited

Blanchard, S. F. 2007. "Recent improvements to the U.S. Geological Survey Streamgaging Program . . . from the National Streamflow Information Program." U.S. Geological Survey Fact Sheet 2007-3080.
Lund, K. D. 2009. Geographer, USGS Iowa Water Science Center, Iowa City office. Personal communication, March 3, 2009.

A. Allen Bradley, Jr.

1 What Causes Floods in Iowa?

ALL RIVERS FLOOD. In fact, they flood with surprising regularity—almost every year or two (Leopold et al. 1964). Some floods are harmless and almost go unnoticed, with water barely spilling out of the river's banks. Others are natural disasters that draw national attention.

Iowa experienced such a disaster not long ago. Floodwaters destroyed homes and businesses, shut down city services, disrupted travel in the region, and damaged farms and cropland. That was 1993. In its aftermath, many Iowans probably assumed they had witnessed the biggest flood that would occur in a very, very long time. And then came 2008, a second major flood disaster in fifteen years.

What do these floods tell us about the nature of flooding in Iowa? This chapter explores some general truths about floods—when they typically occur in Iowa and why. It also examines the common traits that big floods share.

The meteorological and hydrological causes of floods in the Midwest are well known. Floods occur whenever more water runs off the landscape than the river can hold within its banks. Too much runoff is caused by excessive rainfall or melting snow and ice—water amounts far greater than can soak into the ground. If the ground is saturated or frozen and cannot hold more water, it makes flooding that much easier.

Are there certain times of the year when a flood is more likely? The flood history of Iowa (see figure 1-1) shows that spring and summer define the flood season; about 90 percent of all floods occur from March through August. In contrast, floods rarely occur in the fall and winter.

Iowa's climate plays a dominant role in the timing of floods. Fall floods are rare because soils are so dry; by fall, the summer sun and growing plants have depleted the water stored in the ground. Although big rainstorms still do occur,

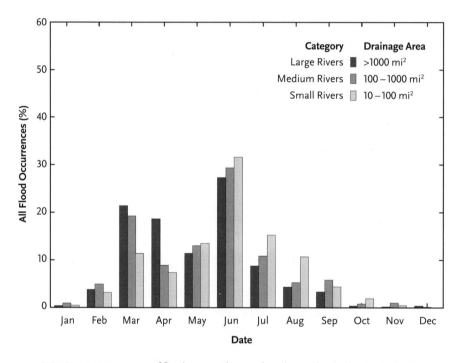

FIGURE 1-1 Occurrences of floods in Iowa by month and according to the river's size (as defined by its drainage area). The flood season is from early spring (March) through summer (August). There is an early springtime peak (March) and a late spring/early summer peak (June) within the flood season. Springtime flooding is more common on larger rivers, while summertime flooding is more common on smaller rivers. *Illustration by the author, based on annual peak discharge data obtained online from the U.S. Geological Survey at waterdata.usgs. gov/nwis. Flood occurrences are for 167 sites in Iowa with at least 30 years or more of operation. Sites along the Mississippi and Missouri Rivers are excluded. For each river, its bankfull discharge is approximated by the annual peak discharge exceeded in two-thirds of the years, which corresponds to a 1.5-year flood recurrence interval (Leopold 1994). A flood occurrence is then defined as an annual peak discharge greater than its bankfull discharge.*

the ground can typically soak up a lot of the rainwater. Winter floods are rare because precipitation is often in the form of snow. While it can rain in winter, and rain on winter's frozen ground can produce flooding, this is uncommon because rainfall intensity is usually much less than at other times of the year. Still, rain and snow during the fall and winter help replenish the moisture in the ground, leaving soils much wetter—and less able to soak up water—at the start of the flood season in spring.

Spring floods occur in a variety of ways. The accumulation of snow during the winter months, and its melt over a relatively short period (of days and weeks), produces many spring floods (Soenksen et al. 1991). Spring weather, characterized by the clash of warm and cold air masses, marks the return of heavy rain with strong low-pressure systems. If widespread rain falls shortly after melting snow has saturated the ground, or if melting is caused by a warm period with rain on snow, a flood is often the result. Such events are responsible for the springtime (March) peak in Iowa floods (see figure 1-1).

Late spring and summer floods are caused by thunderstorms (Soenksen et al. 1991). It is both the heavy rainfall that thunderstorms produce and the location and timing of the rain that determine the magnitude and extent of flooding (see chapter 2). Indeed, most thunderstorms pass too quickly to produce a flood. But certain weather conditions—like a stationary front between warmer and cooler air masses—allow thunderstorms to repeatedly develop and dump heavy rainfall over the same region (Chappell 1986). With the slowly changing weather patterns typical of summer, thunderstorms developing over the span of several days can produce large rainfall accumulations. Such events are responsible for the late spring/early summer (June) peak in Iowa floods (see figure 1-1).

The drainage area of a river (the land area that contributes water to its flow) also plays a significant role in the timing of floods. Springtime flooding is more common on larger rivers, while summertime flooding is more common on smaller rivers and streams (see figure 1-1).

It is a bit of an oversimplification, but in general, larger rivers flood due to a *high volume* of runoff—widespread runoff from snowmelt or rainstorms over an extended period of time. As water runs off the landscape from a large drainage area, it accumulates and concentrates in the river. In contrast, smaller rivers flood due to a *high rate* of runoff—intense rainfall rates from individual thunderstorms can sometimes be enough. Water reaches swales and streams so quickly, and the river rises so rapidly, that flash flooding occurs within hours or a day of the storm.

Although these generalizations about floods describe their common traits, big floods have their own timetable. Iowa's biggest floods are concentrated in the late spring and summer (see figure 1-2); most of the biggest floods have occurred during the months of June and July. The dominance of big floods caused by summertime thunderstorm systems is dramatic for rivers large and small. Although springtime snowmelt and rain floods are very common in Iowa, they simply do not produce the biggest floods we see. Put simply, big floods are different.

Each big flood has its own story, but the floods in 1993 and 2008 stand out in Iowa's flood history. For large and medium-size Iowa rivers, about one in three experienced their biggest flood in 1993 or 2008; for small rivers, about

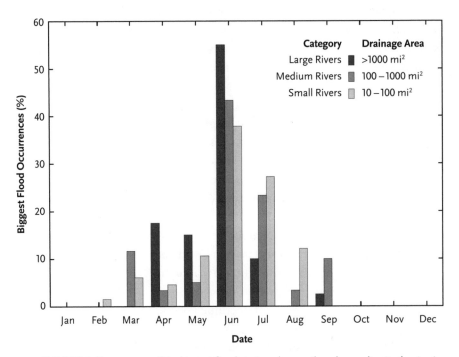

FIGURE 1-2 Occurrences of the biggest floods in Iowa by month and according to the river's size (as defined by its drainage area). The biggest floods are defined by the single biggest flood (peak discharge) on record at each U.S. Geological Survey streamgage site. The biggest floods occur on a different timetable than do other floods (see figure 1-1), typically in late spring and summer. Most were observed in the months of June and July; the June peak accounts for almost half of the biggest floods. Larger rivers tend to have their biggest floods earlier in the year than do smaller rivers. *Illustration by the author, based on annual peak discharge data obtained online from the U.S. Geological Survey at waterdata.usgs.gov/nwis. Results are for 167 streamgage sites in Iowa with at least 30 years of operation. Sites along the Mississippi and Missouri Rivers are excluded.*

one in five experienced their biggest flood then. No other Iowa floods match their widespread impact.[1]

What was so unique about the weather in 1993 and 2008 that created such monster floods in Iowa? In many ways, the precursors to the 1993 and 2008 floods were eerily similar. Both floods occurred after a wet winter and spring. For the eastern third of Iowa, 2008 ranks as the wettest winter in the 114 years that weather records have been kept for the state; this was followed by the second wettest spring on record. Although the winter and spring precipitation were less in 1993, they were still well above average.[2]

Then the rains intensified in the summer. The summer of 1993 stands out as the wettest on record. The precipitation accumulation was a whopping 2.3 times the summer average—an amount only 5.5 inches shy of the average precipitation for a year! Summer 2008 was rainy, with storms producing flooding in June, but precipitation diminished somewhat afterwards; still, it ranks as the ninth wettest summer on record. Looking at precipitation totals from the fall preceding the floods through the following summer, 1993 was the wettest year on record in eastern Iowa, and 2008 was the second wettest.

In both 1993 and 2008, the excessive precipitation over the year produced a large amount of water running off the landscape. Both years were extraordinary in terms of the sheer volume of water coursing through Iowa's rivers. In 1993, the annual discharge volume was the largest seen on record, about three to four times the long-term average at most streamgage sites in the Cedar and Iowa River basins. In 2008, the annual discharge volume was two to three times the average, the second largest on record. Coincidentally, both 1993 and 2008 were preceded by a wetter-than-average year, suggesting that soils may have held more moisture than usual as winter arrived.[3]

Perhaps the most striking difference between the 1993 and 2008 floods was the duration of their summer rainy period. In essence, the 1993 flood was a summer-long event. An unusual and persistent weather pattern caused storms to develop and repeatedly track over the Midwest for nearly two months, from mid-June until mid-August (Wahl et al. 1993). Heavy rainfall accumulations in Iowa (see figure 1-3) resulted from a series of heavy thunderstorms throughout the summer, and affected most of the state at some point. By contrast, the 2008 flood was short-lived; the unusually wet weather persisted for only a few weeks, first dumping heavy rainfall in the northern and northeastern parts of the state, and later encompassing regions farther to the south (see figure 1-3). After heavy rains in late May and early June, Iowa remained relatively dry in the weeks that followed.

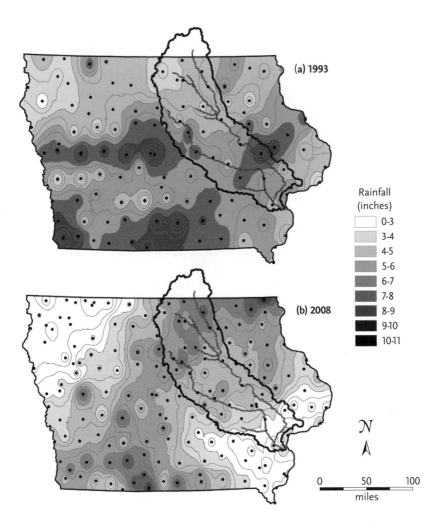

Rainfall
(inches)

	0-3
	3-4
	4-5
	5-6
	6-7
	7-8
	8-9
	9-10
	10-11

FIGURE 1-3 Heavy rainfall accumulations in Iowa during the spring and summer of (a) 1993 and (b) 2008. The maps (with Cedar and Iowa River watersheds outlined in black) show the amount of rainfall accumulation (in inches) for the wettest 5-day period at each weather station (black dot) during the spring and summer. Heavy rainfall amounts were then mapped to give an indication of areas hardest hit by heavy rainfall episodes in each year. In 1993, the largest accumulations were a result of thunderstorms from mid-June through mid-August. Almost two-thirds of the state received heavy rainfall (over 5 inches) in a 5-day period at some point during the summer. In 2008, unusually wet weather persisted for a few weeks, from late May through early June. Although the 2008 accumulations were less statewide, the location and timing of the rain in the Cedar and Iowa River basins played an important role in determining the magnitude of flooding (see chapter 2). *Illustration by Mohamed Habib and the author, using the online precipitation archive from the Iowa Environmental Mesonet at mesonet.agron.iastate.edu, with daily precipitation data from National Weather Service cooperative observer stations.*

The longer rainy period in 1993 produced a much longer flood duration and higher summer runoff volumes, which were clearly felt at the Coralville Reservoir on the Iowa River just upstream from Iowa City (see figure 1-4). In both 1993 and 2008, the Coralville Reservoir filled during the wet spring, as reservoir releases were managed (according to its operating guidelines) to reduce flooding downstream. In June 2008, the rising water in the reservoir began overflowing the emergency spillway, but reservoir levels quickly lowered with the drier weather after the flood crest. In July 1993, after water began overflowing the emergency spillway, reservoir levels remained near capacity until September.

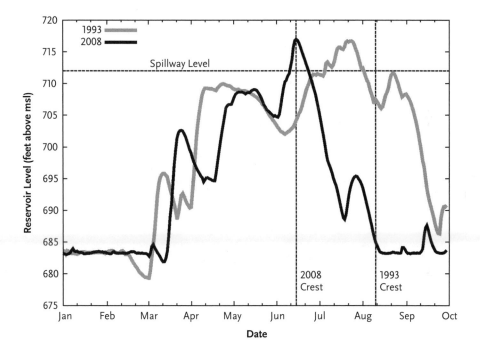

FIGURE 1-4 Coralville Reservoir elevation hydrographs in feet above mean sea level (msl) for the flood years of 1993 and 2008. Water begins to flow over the dam's emergency spillway when the reservoir reaches the spillway elevation (712 feet msl; see chapter 10). The dates when the Iowa River at Iowa City crested during the 1993 and 2008 floods are indicated by vertical lines. In both flood years, the Coralville Reservoir stored water during the spring months in an attempt to reduce downstream flooding. Thus, starting in March, the reservoir level rose above the target operating level of 683 feet msl. In 2008, reservoir levels quickly lowered after the flood crest and returned to target operating levels in August. In 1993, the longer duration of the flood and its greater runoff volume kept reservoir levels near capacity into September. *Illustration by the author, based on daily reservoir level data obtained online from the U.S. Army Corps of Engineers at rivergages.com.*

The differences in flood duration are also evident in the highest river levels seen at different locations (see figure 1-5). In 2008, the floods crested throughout Iowa from late May to mid-June. Afterward, they receded through the remainder of the summer. But in 1993, rivers crested over a span of five months and remained high for much of the summer.

The two flood events also differed significantly in the size of the area they affected. Overall, the impact of the 1993 flooding was significantly more widespread. Flooding affected not only Iowa but also portions of Illinois, Kansas, Minnesota, Missouri, Nebraska, North Dakota, South Dakota, and Wisconsin (see figure 1-6). That flood ranks among the biggest in both the upper Mississippi and lower Missouri River valleys. In contrast, the impact of the 2008 flooding was most severe in Iowa and Wisconsin, although the flooding also affected portions of Illinois, Indiana, Minnesota, Missouri, Nebraska, and South Dakota. On the Mississippi River, the 2008 flood approached 1993 levels only at the streamgage in Keokuk, Iowa; magnitudes diminished rapidly downstream.

Do these two floods provide clues as to what the next big flood may be like? They do suggest that it most likely would be a late spring or summertime flood, since this is the time when persistent and widespread thunderstorms can produce large rainfall accumulations (Hirschboeck 1991). If 1993 and 2008 are any indication, what happens in the winter and spring is important in setting the stage for a big flood. In both those years, after a wet winter, snowmelt

FIGURE 1-5 Flow hydrographs in cubic feet per second (cfs) for the flood years of 1993 and 2008 for (a) the Cedar River at Cedar Rapids, and (b) the Iowa River at Iowa City. The Cedar River is a free-flowing river, with no significant water storage upstream of Cedar Rapids. As water runs off after rains and snowmelt, it travels unimpeded, and the river level rises and falls with the flow's passage. In contrast, the Iowa River is dammed by the Coralville Reservoir upstream of Iowa City; the releases from the reservoir are managed to reduce downstream flooding. As water was being stored in the Coralville Reservoir during the spring of 1993 and 2008, operational releases remained near 10,000 cfs until the reservoir's capacity was exhausted. In 2008, the two rivers crested in mid-June and then receded. In 1993, the two rivers crested 129 days apart. At Cedar Rapids, the Cedar River crested near the beginning of 1993's spring flooding on April 4. River levels remained high, but the heavy summer rains never quite pushed the river past its springtime crest. The 1993 crest in Iowa City on August 10 was at the end of the summer-long event. Then, with Coralville Reservoir near capacity and its release still high, locally heavy rains and tributary runoff from Clear Creek and Rapid Creek, which enter the Iowa River between the reservoir and Iowa City, pushed the river to its highest level of the summer. *Illustration by the author, based on instantaneous and daily discharge data obtained online from the U.S. Geological Survey at waterdata.usgs.gov/nwis. Flood stage as determined by National Weather Service Advanced Hydrologic Prediction Service, http://www.weather.gov/oh/ahps/.*

(a) Cedar River at Cedar Rapids

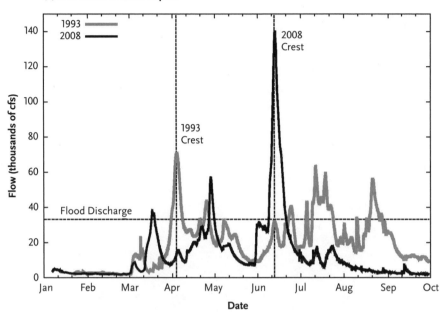

(b) Iowa River at Iowa City

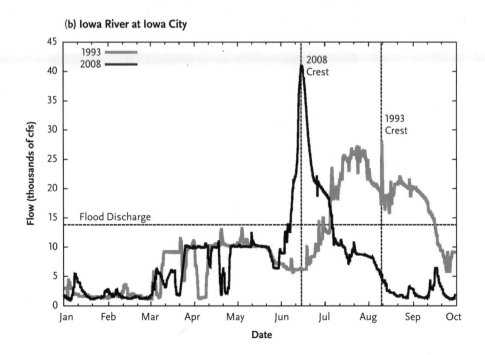

and spring rains on then-saturated soils conspired to push rivers out of their banks in early spring throughout much of the state. As summer approached, the landscape could not soak up much more water, and rivers were primed to flood again when heavy rains ensued. In the future, a wet and active spring flood season should send a signal that the worst may not be over.

Are there any lessons that we should learn from witnessing two major floods in 15 years? One such lesson is that flood disasters are to be expected, and some

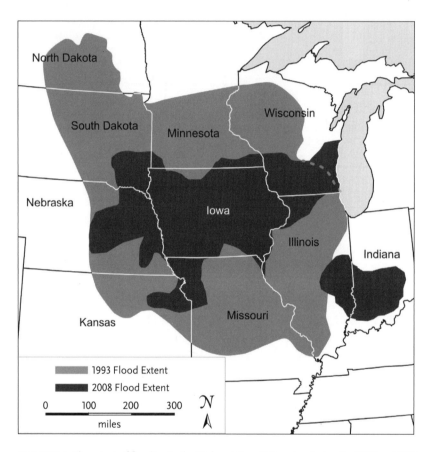

FIGURE 1-6 The extent of flooding in the Midwest Corn Belt region during the 1993 and 2008 floods. Overall, the flooding in the summer of 1993 was more widespread. It affected a larger area and two major rivers, the upper Mississippi and the lower Missouri. The flooding from late May to mid-June in 2008 was most severe in Iowa and Wisconsin, although storms over the period produced areas of flooding across the region. *Illustration by Radoslaw Goska, with 1993 flood area as in Parrett et al. (1993) and 2008 flood area based on the author's analysis of peak discharge data at U.S. Geological Survey streamgages. Both are approximations of the areas with flooding streams.*

may be even bigger than the ones we have already seen. We do indeed need to be very concerned about future flooding. Just because we have experienced a flood does not mean that the risk of another is less in subsequent years. Floods are random events; as any gambler knows, a run of bad luck can (and does) occur by random chance.

Unfortunately, many first perceived the 1993 flood as a once-in-a-lifetime event, an unsurpassable benchmark against which all other floods would be measured—a perception fueled by its long-lived nature, its widespread impact, and the media attention surrounding the disaster. The 1993 flood was unprecedented in many ways, but it was not extraordinary at all locations in the Midwest where flooding occurred (Parrett et al. 1993). Regrettably, the mistaken belief that 1993 was a benchmark may have shaped attitudes about the need to prepare for future big floods, and the response (or lack of response) to the rising floodwaters in 2008. Will we make the same mistake in the aftermath of the Iowa floods of 2008?

Notes

1. These Iowa flood facts are based on the author's analysis of annual maximum flood discharge records at streamgage sites, obtained online from the U.S. Geological Survey (http://waterdata.usgs.gov/nwis).

2. The precipitation facts for the 1993 and 2008 floods are based on my analysis of climate division precipitation data for Iowa, obtained from the National Climatic Data Center (www.ncdc.noaa.gov). The precipitation data span a 114-year period (1895–2008) for the three climate divisions covering the eastern third of Iowa.

3. The analysis of annual river discharge volumes for the 1993 and 2008 floods is based on streamflow information for Cedar and Iowa Rivers obtained online from the U.S. Geological Survey (http://waterdata.usgs.gov/nwis).

References Cited

Chappell, C. F. 1986. "Quasi-stationary convective events." In *Mesoscale Meteorology and Forecasting*, ed. P. S. Ray. Boston: American Meteorological Society.

Hirschboeck, K. K. 1991. "Climate and floods." In *National Water Summary 1988–89: Hydrology Events and Floods and Droughts*, compiled by R. W. Paulson, E. B. Chase, R. S. Roberts, and D. W. Moody, pp. 67–88. U.S. Geological Survey Water Supply Paper 2375. http://pubs.er.usgs.gov/usgspubs/wsp/wsp2375.

Leopold, L. B. 1994. *A View of the River*. Cambridge, Mass.: Harvard University Press.

Leopold, L. B., M. G. Wolman, and J. P. Miller. 1964. *Fluvial Processes in Geomorphology*. San Francisco: W. H. Freeman and Co.

Parrett, C., N. B. Melcher, and R. W. James, Jr. 1993. "Flood discharges in the Upper

Mississippi River Basin, 1993." In *Floods in the Upper Mississippi River Basin 1993*. Reston, Va.: U.S. Geological Survey Circular 1120-A.

Soenksen, P. J., D. A. Eash, H. J. Hillaker, and E. M. Gordon. 1991. "Iowa floods and droughts." In *National Water Summary 1988-89: Hydrology Events and Floods and Droughts*, compiled by R. W. Paulson, E. B. Chase, R. S. Roberts, and D. W. Moody, pp. 279–286. U.S. Geological Survey Water Supply Paper 2375. http://pubs. er.usgs.gov/usgspubs/wsp/wsp2375.

Wahl, K. L., K. C. Vining, and G. J. Wiche. 1993. "Precipitation in the Upper Mississippi River Basin, January 1 through July 31, 1993." In *Floods in the Upper Mississippi River Basin 1993*, Reston, Va.: U.S. Geological Survey Circular 1120-B.

Witold F. Krajewski
Ricardo Mantilla

2 Why Were the 2008 Floods So Large?

In June 2008, eastern Iowa experienced some of the worst flooding ever recorded. Floods, commonly defined as river waters overflowing their banks, devastated cities and the countryside alike. Some 1.2 million acres of Iowa's agricultural land was affected by floodwaters.[1] From a plane, it was difficult to decipher the main channel of many rivers. The confluence of the Iowa and Mississippi Rivers seemed like a sea dotted with silos that protruded from the water's surface. In Cedar Rapids, a six-foot-tall man standing on the west bank levees would have had water flowing six feet above his head.

The peak flow on the Cedar River at Cedar Rapids reached 140,000 cubic feet per second (cfs) on June 13. This extremely large flow was a full five times as large as the river's average annual peak flow at this site.[2] On the Iowa River at Marengo, the peak flow reached 51,000 cfs on June 12, four times as large as the average annual peak flow at this site. Downstream from Marengo, the Coralville Reservoir offered some protection, but in Iowa City (with a downtown area about 8 river miles below the dam), the Iowa River still reached a peak flow of 41,100 cfs and flooded significant areas of the University of Iowa campus. Peak flow there was three times the average annual peak flow (USGS 2009).

The sense of the 2008 floods' tremendous magnitude also is portrayed in figures 2-1 and 2-2, hydrographs that plot May and June discharge (river flow) at several streamgages on the Cedar and Iowa Rivers and their tributaries. Note how in most instances, especially for rivers draining large areas, the 2008 peak greatly exceeded the average annual peak flow, a level that roughly corresponds to a river filling its banks to the brim.

What was the genesis of these floods? Why were they so large? Many have pondered these questions, trying to grasp unique features of weather and land-scape that might provide answers. Here we consider three contributing factors: the severe winter that preceded the floods, the high-intensity rainstorms of late May and early June, and the possibility of a perfect storm—not necessarily extremely large, but one in which precipitation was perfectly timed and located to raise the flow in river drainage networks to extraordinary levels.

First, consider the preceding season. The region experienced one of the snowiest winters in recent memory. Average snow depth across Iowa reached 11 inches by late February (NOHRSC 2008). The snowmelt saturated the ground, which remained wet well into May and significantly delayed the planting of crops in many parts of the state. With cool temperatures and croplands remaining bare, little vegetation was available to dry the soil through transpiration (the pulling of moisture from plant roots through leaves into the air).

Could these winter conditions help explain the June flooding? Examine the graphs in figure 2-3 (also plate 14). Panel A shows the daily variation of snow cover, snow depth, and snow water equivalent (the amount of water that would result from melting the snow) for December 2007 through March 2008, averaged over the state of Iowa (NOHRSC 2008). Note that while average snow depth reached over 11 inches by the end of February, the snow was virtually gone by the end of March, with snowmelt hastened by March temperatures rising above freezing (panel B; IEM 2008). The effect of this snowmelt on March flooding is clearly seen in panels D and E. That month, river discharges at both Cedar Rapids and Marengo rose in a single wave, reaching nearly 40,000 cfs and 18,000 cfs respectively, but then dropped back to normal levels by the end of March (USGS 2009). Thus, the heavy winter snows were not directly responsible for later flooding.

The snows did, however, lead to flood-prone conditions. Some of the melting snow remained in the fields as soil moisture. Thus, when the April rainstorms arrived (panel C; IEM 2009), the soils could absorb little of the spring rainfall. Instead, the wet ground produced significant runoff and additional high dis-charge waves in the rivers (note April discharges in panels D and E). Flooding from these storms peaked around the end of April.

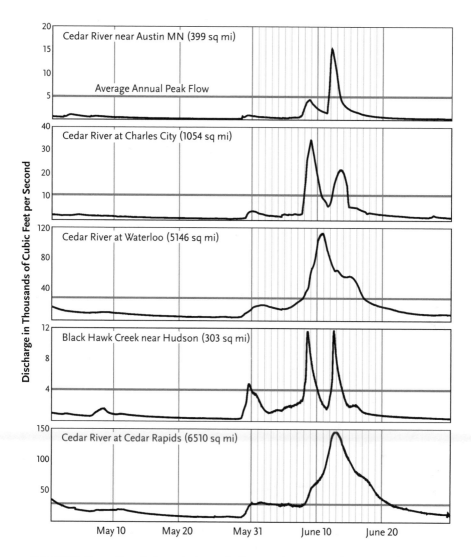

FIGURE 2-1 Streamflow discharge hydrographs for several streamgage locations in the Cedar River basin through the 2008 flood period, displaying flow phenomena explained in this chapter. Locations of Iowa streamgages are shown in plate 1 and figure 2-5; Black Hawk Creek is a tributary that enters the Cedar River at Waterloo. Vertical lines, May 31–June 20, delineate individual days. Upstream drainage area in square miles follows the station name. These hydrographs show that the discharges rose throughout the basin during the flood, although to different degrees (note that the discharge scale on the left is different for every location shown): the discharges increase for locations draining larger and larger areas, with the small creek (Black Hawk Creek) showing a smaller but very sharp rise and fall. River discharges such as those displayed here reflect both precipitation and basin geomorphology. *Illustration by Radoslaw Goska, based on USGS 2009.*

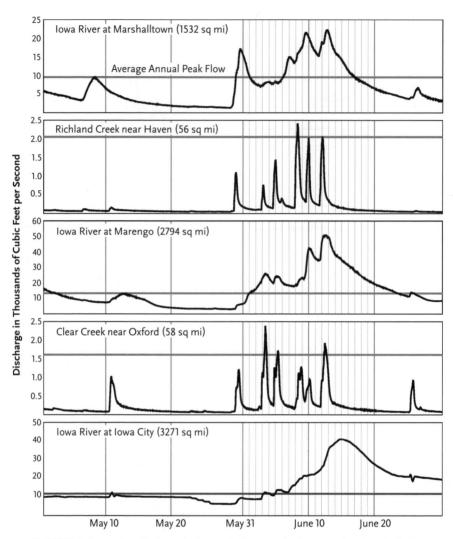

FIGURE 2-2 Streamflow discharge hydrographs for several streamgage locations in the Iowa River basin through the 2008 flood period, displaying flow phenomena that are explained in this chapter. Streamgage locations on the Iowa River are shown in plate 1 and figure 2-5; Richland Creek is a tributary that enters the Iowa River between Tama and Marengo, and Clear Creek enters the Iowa River in Coralville. Vertical lines, May 31–June 20, delineate individual days. Upstream drainage area in square miles follows the station name. These hydrographs show that the discharges rose throughout the basin during the flood, although to different degrees (note that the discharge scale on the left is different for every location shown). The tributaries, with their smaller drainage basins, demonstrated sudden discharge rises and falls (rather than slower, more sustained changes), and the Coralville Reservoir and Dam (between Marengo and Iowa City) attenuated peak flows in Iowa City. River discharges such as those displayed here reflect precipitation and basin geomorphology, as well as the effects of the reservoir. *Illustration by Radoslaw Goska, based on USGS 2009.*

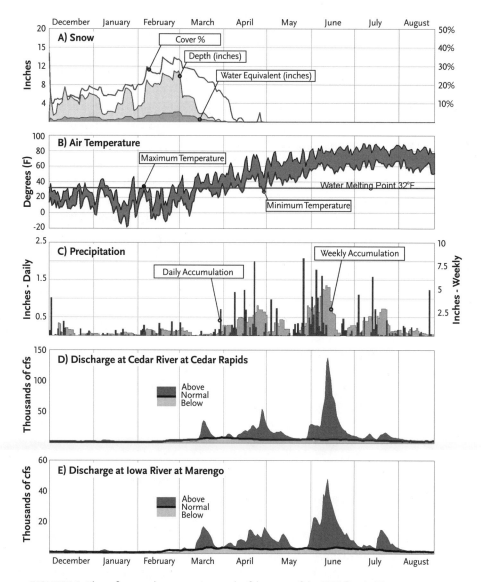

FIGURE 2-3 These five panels summarize much of the story of the 2008 floods. The unusually wet winter deposited an average of 11 inches of snow across the state (panel A, based on daily records from NOHRSC 2008). When air temperatures rose above freezing (panel B, temperatures above Iowa and Cedar River basins, IEM 2008), the snow melted rapidly, causing small floods in March (panels D and E, based on USGS 2009). However, heavy precipitation (panel C, rainfall over Cedar and Iowa River basins, IEM 2008) falling on saturated soils quickly led to additional flooding, first in late April and then spectacularly in June. Note that panel E shows the river discharge above Marengo rather than at Iowa City, because river flows at Iowa City are managed by the Coralville Dam. Plate 14 is a color version of this graphic. *Illustration by Radoslaw Goska.*

For most of May, Iowa received relatively little rainfall. River levels through-out eastern Iowa returned to normal by May 25, as if providing the proverbial calm before the storm. Farmers planted crops late in May, and air temperatures reached typical levels for this time of the year.

And then came the rains. In late May and early June, a series of severe storms rolled across the state, affecting different regions at different times, but generally dumping water on the already wet land. Conditions were prime for June's flooding and devastation. This situation was reminiscent of June 1993, when frequent storms resulted in a record flood over the entire Upper Mississippi River basin.

Several media outlets and government agencies spoke about rainfall records being broken throughout the state (NCDC 2008), convincing Iowans that the subsequent epic floods were a mere consequence of never-before-seen storms. However, while the high-intensity storms were the source of June's massive flood wave, a careful examination of the water volumes dumped over the state reveals that causation of flooding was more complex.

Figure 2-4 shows the magnitude of the storms from late May through June at over 100 weather stations throughout the state. To put these storms into historical perspective, we compared the maximum rainfall accumulation over one day, one week, two weeks, and three weeks at these stations (IEM 2008) to Iowa's long-term mean annual maximum precipitation (MAMP[3]) for the same duration of time. This comparison serves as a reference point: a rainfall close to the MAMP would not be considered rare, but a rainfall much larger than the MAMP would indeed be unusual.

The results show that when the one-day maximum rainfall at each station in June 2008 is compared to its MAMP (upper left map in figure 2-4), the rainfall at about half the stations was actually less than the MAMP (the ratio is less than one). By this standard, the individual storms that moved across Iowa in June were fairly common; no single epic storm preceded the floods. On the other hand, over longer time periods (the remaining three maps in figure 2-4), the late May/June 2008 maximum rainfall accumulations were much more extreme, exceeding their MAMP over most of Iowa (that is, the ratio is much greater than one).

Though it is tempting to attribute the unprecedented nature of the 2008 floods to the heavy rainfall over a week or more, this is not the whole story. Large rainfall accumulations were seen throughout Iowa in 2008, not just at locations in eastern Iowa where the worst flooding occurred. Furthermore, the historical records for weather stations show that even larger rainfall accumulations have

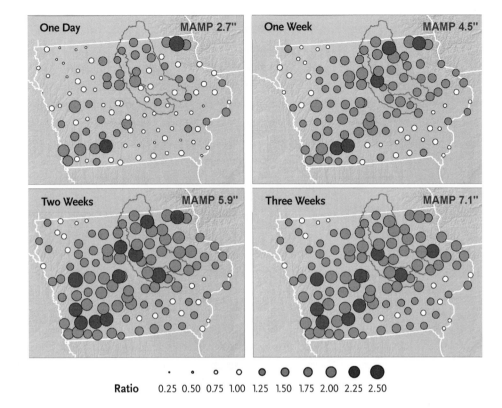

Ratio	0.25	0.50	0.75	1.00	1.25	1.50	1.75	2.00	2.25	2.50

FIGURE 2-4 These maps relate the largest rainfall accumulation observed during the 2008 floods for different time periods in late May and June to the mean annual maximum precipitation (MAMP, see endnote 3) at each of Iowa's NWS Cooperative Observer weather stations. Circle size is proportional to the value of the ratio, and the three circle colors—white, gray, and black—show locations with ratios between 0 and 1, 1 and 2, and greater than 2 respectively. The Cedar and Iowa River basins upstream from Cedar Rapids and Marengo are outlined as in figures 2-5 and 2-6 (see plate 1 for delineation of the full watersheds). The MAMP is the average of the largest storm on record for each year, extended throughout the lifetime of the particular weather station, usually 50–100 years. The MAMP for the entire state is 2.7 inches (for a one-day period), and it increases to 7.1 inches (for a three-week period, with the MAMP being calculated for each time period in question, using the mean annual maximums for each year and for that period). On these maps, smaller circles (with precipitation ratios under 1) indicate that the 2008 rainfall at a given site was less than the MAMP—and thus was not extraordinary in any way. For the one-day map, about 50% of the 2008 one-day rains were less than the MAMP. By this standard, individual single-day storms that moved across Iowa in June were fairly common. When considering time periods of one, two, or three weeks, the late May/June 2008 rainfalls exceeded the MAMP more frequently (ratios were greater than one), but still 10 percent or fewer locations measured rainfall accumulations greater than twice the state's mean annual maximum. Because such rainfalls are not the most extreme that Iowa has experienced in the past, the floods of 2008 cannot be explained solely on the basis of extraordinary rainfalls. *Illustration by Radoslaw Goska, based on IEM 2008.*

occurred in other years, so the 2008 rains were not the worst that Iowa has ever seen (while the floods were the worst at many locations).

So why were the floods so severe on the Iowa and Cedar Rivers? How then is it possible that a series of rainstorms that deposited around twice the annual average peak rainfall volume produced these 2008 super floods, which were four or five times larger than the average peak flow in eastern Iowa?

Perhaps the culprit was a perfect storm, where the timing and location of rains conspired to maximize flood intensity at the hardest-hit locations. To understand this concept, one needs to follow the flow of water from where it falls to a specific point on the river. River basins or watersheds are defined by all the points in a landscape that drain into a common point on a river. Every major basin holds thousands of streams and creeks that form a complex river network that drains the landscape. The speed of water's flow through the interconnected basin depends on several factors, most importantly land cover, soil type, level of soil saturation, proximity to a drainage channel, and the slope and flow resistance along the channel. Dry ground can absorb rainfall, which then travels slowly underground. But rain that falls on the surface of saturated ground runs quickly off the surface and into the closest drainage channel. Once in a channel, water flows even more rapidly, but rarely faster than 5 feet per second in Iowa rivers.

For any location in the basin, we can roughly estimate the time that it takes for water in the river network to reach the basin outlet. Figure 2-5 (also plate 12) shows the Cedar and Iowa River basins above Cedar Rapids and Marengo (the Coralville Reservoir), with their zones of water travel time. With this map in mind, you can imagine that the water you watch at one point on a river is the sum of the runoff generated hours or days earlier from multiple upstream locations. Now imagine that when a single storm hits Charles City, Cedar Rapids residents will observe its stormwater flow about four days later. If a second storm hits Waterloo a day or two after the Charles City storm, when that stormwater is flowing through Waterloo, the water from both storms will combine to arrive in Cedar Rapids about the same time, creating a water traffic jam analogous to rush hour traffic jams.

Figure 2-6 (also plate 13) demonstrates that this is indeed what happened in 2008: The rainstorms that occurred in early June created a massive water traffic jam that resulted in the worst flooding in Cedar Rapids history. Rain in the upper watershed falling on June 8 moved downstream to combine with rain falling lower in the watershed on June 12, the consecutive storms compounding to produce an unexpected, rapid river rise and single well-defined and extremely large peak flow in Cedar Rapids on June 13. Our analysis is based on data from the U.S. National Weather Service's NEXRAD weather radars, which were deployed in 1990. These radars allow us to track storms and rainfall for the first time, providing an unprecedented picture of the location and timing of daily rainfall accumulations.

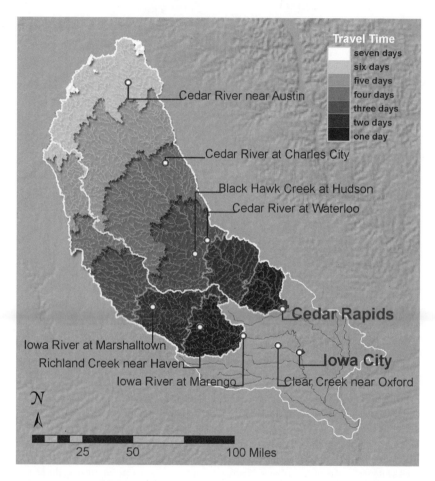

FIGURE 2-5 Map of the average travel time required for streamwater to flow through the drainage network of the Cedar and Iowa Rivers, down to Cedar Rapids and Marengo (just above the Coralville Reservoir on the Iowa River), with the entire watershed outlined in white. For the Iowa River, we depict locations only down to the Coralville Reservoir because its human-managed operation influences travel times lower on the Iowa River. See figure 2-6 for the significance of this travel-time map, plate 1 for locations of Iowa and Cedar River streamgage sites, and figure I-2 for the broader location of the watersheds. Plate 12 is a color version of this graphic. *Illustration by Radoslaw Goska, based on Mantilla et al. 2006.*

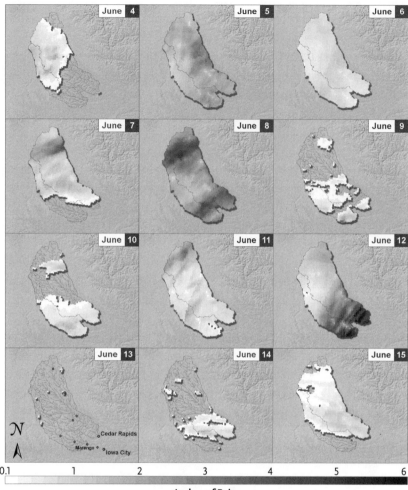

Inches of Rain

FIGURE 2-6 These maps show the amount of rain (indicated by shading) that fell on the Cedar and Iowa River basins upstream from Cedar Rapids and Marengo throughout the June 2008 flood period. (See figure 1-2 to locate these watersheds.) With this knowledge of the location and timing of daily rainfall accumulations and with use of the figure 2-5 travel-time map, we can explain the 2008 flooding of Cedar Rapids in this way: The lower band of the June 8 intense rainfall reached Cedar Rapids a day or two after that storm, creating the first rise of the big flood (figure 2-1, bottom panel). The second, more northerly band of June 8 rainfall required a longer travel time, about five days, and arrived at Cedar Rapids together with the massive rainfall that fell just north of Cedar Rapids on June 12. The compounding of these consecutive storms produced the unexpected, rapid river rise that caused the single, well-defined, and extremely large peak flow in Cedar Rapids and resulted in the worst flooding in Cedar Rapids' history. Small circles on the June 13 map are streamgage stations. Plate 13 is a color version of this graphic. *Illustration by Radoslaw Goska, based on U.S. National Weather Service NEXRAD data, from Hydro-NEXRAD.net.*

This same concept also explains why small basins experienced only mild flooding. Because these basins typically drain stormwaters in a single day, consecutive storms usually do not cause a water traffic jam. To visualize this process, turn again to figure 2-2. Notice how Richland Creek and Clear Creek (see locations in figure 2-5), with basins under 60 square miles in area, responded to each early June storm with a single short-lived peak. Because peak rainfall accumulations over a one-day time period were not unusually large, peak floods observed were not extreme in smaller basins.

A similar analysis holds for the peak flows observed in the Iowa River basin, except for points downstream from the Coralville Reservoir (just upstream from Iowa City), which regulates the river's flow (see figure 2-2). Reservoirs reduce downstream flooding by intercepting flood waves and storing water temporarily, releasing it at a lower rate over an extended period of time. However, if inflows fill the reservoir, its storage is depleted and the dam's emergency spillway will eventually be overtopped, an event that happened with the Coralville Reservoir on June 10. At this time, the dam lost much of its ability to regulate downstream flows and flooding.

Once the spillway was overtopped, the reservoir's large discharge caused additional problems downstream, where the Iowa River's flow was high enough to prevent its tributary Clear Creek from discharging its own flood flows into the Iowa River. Clear Creek enters the Iowa River about 5 river miles downstream from the dam, in Coralville. There, rather than receiving and moving Clear Creek's discharge, the Iowa River backed up into Clear Creek. The combined waters from these two rivers flooded Coralville's business district.

In summary, the story of the June 2008 floods in eastern Iowa includes a complexity of processes. More precise evaluation of various contributions to the flooding requires additional analysis. Interesting questions to examine include the exact role of the spring's high soil moisture, and the possibility that Iowa's June storms were remnants of an oceanic tropical cyclone near Mexico that was pulled into the Midwest by the Jet Stream.

River basins, unlike many human-made engineering systems, cannot be taken into a laboratory or studied under a microscope. Investigations require large amounts of data, significant computational resources, and a comprehensive background in many fields of science and engineering. Thorough understanding of past events is a prerequisite to preparation for the future. Only science-based guidelines can effectively address inevitable future floods. Research leading to better understanding of the exact roles of snowfall, rainfall, and the river network as a conveyor belt will greatly enhance the success of flood mitigation efforts.

Acknowledgments

We appreciate the assistance of Radoslaw Goska in preparing the figures. Research outlined in this chapter is being supported by National Science Foundation awards CBET-0842682 and EAR-0844101 and by the Rose and Joseph Summers endowment.

Notes

1. This estimate was made using data on harvesting intention change from NASS 2008.

2. The average annual peak flow is calculated by taking the maximum flow observed in each year on record. For the Cedar River at Cedar Rapids, the historical record goes back more than 100 years. This quantity is also defined as the mean annual flood by the United States Geological Survey (USGS). For USGS definitions, see http://water.usgs.gov/wsc/glossary.html.

3. The MAMP is calculated from a set of rainfall accumulation time series with a specific accumulation interval (e.g., one day, one week, two weeks, etc.). The sample to calculate the average is formed by the annual maximum for each year and for each streamgage. Thus if I have 100 stations, each with 100 years of record, the MAMP is the average of 10,000 values.

References Cited

Iowa Environmental Mesonet (IEM). 2008. Ames, Iowa: Iowa State University Department of Agronomy. http://mesonet.agron.iastate.edu/.

Mantilla, R., V. K. Gupta, and O. Mesa. 2006. "Role of coupled flow dynamics and real network structures on Hortonian scaling of peak flows." *Journal of Hydrology* 322 (1–4): 155–167.

National Agricultural Statistics Service (NASS). 2008. "June 2008 acreage report." Washington, D.C.: U.S. Department of Agriculture. http://usda.mannlib.cornell.edu/usda/nass/Acre/2000s/2008/Acre-06-30-2008.pdf.

National Climatic Data Center (NCDC). 2008. "Climate of 2008. Midwestern U.S. flood overview." Asheville, N.C.: U.S. Department of Commerce. http://www.ncdc.noaa.gov/oa/climate/research/2008/flood08.html.

National Operational Hydrologic Remote Sensing Center (NOHRSC). 2008. "The National Operational Hydrologic Remote Sensing Center." Asheville, N.C.: National Weather Service, National Oceanic and Atmospheric Administration. http://www.nohrsc.nws.gov/.

U.S. Geological Survey (USGS). 2009. *Water-Resources Data for the United States, Water Year 2008*: U.S. Geological Survey Water-Data Report WDR-US-2008, Iowa sites. http://wdr.water.usgs.gov/wy2008/search.jsp.

Richard A. Fosse

3 Iowa City and the Flood

IN TIMES OF FLOODING, communities engage in a flurry of activities that test the limits of everyone involved. Extreme demands and rapidly changing circumstances may produce chaotic conditions that hamper a community's ability to respond. Advance planning and focus are crucial if the government is to carry out its primary responsibilities to its citizens: protect the safety of residents, safeguard critical infrastructure, share information with the public, and where possible, provide resources to assist residents in their preparations for the flood. Emergency responses need to be planned and practiced beforehand. Such proactive efforts are particularly important in a multi-jurisdictional metropolitan area such as ours. Iowa City, Coralville, the University of Iowa, and Johnson County are bound together by an interdependence created by common threats and limited resources. Our success in dealing with floods and other emergencies relies on our ability to communicate, coordinate, and cooperate with one another and with the public.

Outlined below is Iowa City's response to the flood of 2008, framed around the events as they unfolded. Similar responses were simultaneously made in Coralville, Johnson County, and the University of Iowa; specific efforts of these other groups are identified throughout the chapter.

Tuesday, June 3: Our response began with a phone call from the U.S. Army Corps of Engineers (Corps). The Corps operates the Coralville Reservoir, formed by a large dam on the Iowa River approximately eight river miles upstream from Iowa City's downtown area (see figure I-4 for this and other site locations). The Corps' call confirmed our concerns that river conditions and projected weather patterns were likely to cause flooding.

Wednesday, June 4: Although the future remained unclear, we began to prepare for a significant flood event. Because we lie just downstream from a dam that regulates the release of water, our floods are effectively set in slow motion. This gives us more time to strategize and contemplate our options. However, good strategies rely on good information, which makes communication one of our most important tools. On June 4, we had our first conference call with the Corps, establishing a daily ritual that continued throughout the flood. Calls were facilitated by the University of Iowa and included all local entities, ensuring that everyone started each day with the best and the same information.

We also started getting information out to the public through daily press releases and a page on our city's website dedicated solely to flood information. We included links to Corps and U.S. Geological Survey (USGS) websites, which contained the same real-time data for reservoir and river levels that we were using.

Thursday, June 5: Sandbagging was now underway in a number of residential and commercial areas. Without proper technique, sandbagging is rarely successful and can be a waste of resources. To address this, we sent staff to key locations to provide guidance, to arrange delivery of necessary resources, and to coordinate volunteers. We updated our website to provide links demonstrating proper sandbagging techniques.

Our flood response was expanding rapidly and now included many city departments and divisions. We reached a point at which it was necessary to activate our Emergency Command Post and set up an organizational structure tailored to emergency management.

Johnson County reached the same conclusion and opened its Emergency Operations Center to facilitate interagency actions and to channel resources. Outside organizations such as the National Guard, Iowa Department of Transportation, Red Cross, and Salvation Army were all deployed through this center. The County Emergency Operations Center also allowed the intertwined communities to communicate and coordinate as needed.

The model used to restructure our organizations during the flood was developed by the Department of Homeland Security and is called the National

Incident Management System (NIMS). NIMS was developed after September 11, 2001, to address coordinating different organizations during a single major disaster. This was our first disaster that fully utilized the NIMS model, and it notably improved our ability to organize, communicate, and respond.

Friday, June 6: The Corps projected that water might flow over the reservoir's emergency spillway, possibly as soon as June 10. This drove home the fact that we were facing more than a moderate flood. We ramped up our outreach by establishing a call center to answer questions and match volunteers with homes and businesses needing assistance. Over the next two and a half weeks, the call center handled 8,656 calls.

We had concerns that not everyone at risk understood the seriousness of the situation, so we held a press conference to emphasize to residents that it was time to begin preparations, if they had not done so already.

Monday, June 9: The Corps was now projecting a peak discharge of 28,000 cubic feet per second (cfs)[1] on June 17, a peak roughly equal to that of the flood of 1993.

Tuesday, June 10: Heavy rains continued and the Corps revised its projections to a peak flow of 32,000 cfs on June 16, significantly higher than that of the flood of 1993. We were now moving into uncharted territory and could no longer rely on experience to guide us. We had to predict problems that might occur and where possible take steps to limit damage. For example, we now expected water to flow over the deck of the Park Road Bridge, something never before seen. The rising water would create buoyant forces that could lift the bridge deck and push it off its piers. Besides destroying the bridge, such an event might create a temporary obstruction that could cause 1 to 3 feet of additional flooding upstream, followed by a surge of water downstream. To reduce this risk, we drilled holes in the bridge deck to allow air to escape and reduce the buoyant forces. It worked exactly as planned and air spouted from the holes as water overtopped the bridge, releasing significant pressure from under the bridge deck (figure 3-1). The bridge survived the flood (figure 3-2). Johnson County utilized this same technique on a number of its bridges with similar results.

Wednesday, June 11: The Corps raised its peak flow projection slightly to 33,000 cubic feet per second (cfs), but what worried us more was the potential for additional rainfall from storms projected to move through the state in the late evening and early morning hours. Sandbagging efforts intensified.

Thursday, June 12: Although not the peak of the flood, this 24-hour period was the most intense. The rains moved in shortly after midnight. In addition,

FIGURE 3-1 Air bubbles spout up through the water from holes drilled in the Park Road Bridge deck. Holes were drilled in the bridge deck prior to the flood to allow trapped air to escape from under the bridge. Trapped air would have created an upward lift on the bridge deck that could have caused the bridge to be washed off its piers. *Photograph by Bob Hardy, city of Iowa City.*

FIGURE 3-2 A total of fifty-two tons of floating debris was removed from the upstream side of the Park Road Bridge during the flood. Floating debris posed a number of concerns, ranging from structural damage caused by impact to environmental threats from the unknown contents of orphaned containers. *Photograph by Denny Gannon, city of Iowa City.*

the online data indicated that the Iowa River was rising faster than expected throughout the night and it appeared that the sandbag levee protecting the Parkview Terrace neighborhood would overtop well before daylight. Mandatory evacuations were implemented about 2 A.M. for residents closest to the levee. As it turned out, the levee held until midmorning, but when the area flooded it cut off the single road to the neighborhood, making evacuation of the remaining residents difficult.

In the morning, the Corps reported that the widespread storm had produced significant rainfall in the Iowa River basin, and they revised the projected flood peak to 40,000 cfs on June 17. This dramatic increase changed everything. Critical infrastructure, such as drinking water sources and wastewater facilities, was now in real jeopardy and protecting it became our top priority. With three bridges already closed it became apparent that our city could be split by the closure of the remaining bridges. All city services—everything from fire protection to garbage collection—prepared for independent east- and west-side operations by staging equipment on both sides of the river. Engineers were posted 24 hours a day at our remaining bridges to monitor their condition and safety.

Preparing critical infrastructure for the rising water required strategic decisions about what was feasible to protect. We directed our limited resources to those facilities that we believed we could save and we worked to limit damage to facilities likely to be flooded. For example, based on the new peak flow projections, our north wastewater treatment plant would be completely flooded. Our staff quickly devised a plan to protect key components of the facility, and everything else was shut down to limit damage to electrical systems. Chemicals were removed or secured and other steps were taken to minimize submersion damage. The Coralville Wastewater Department staff took similar actions by working in several feet of floodwater to remove control panels and other expensive items from four stormwater lift stations and two sanitary sewer lift stations. These combined efforts saved hundreds of thousands of dollars and, perhaps more importantly, accommodated a quick restart of city services after the flood.

During the mid-afternoon, water broke through the railroad embankment that had served as a de facto levee and protected much of Coralville's commercial area along Highway 6 (figure 3-3). This triggered a scramble to retrieve temporary pumps, a crucial flood resource, stationed in the area.

Friday, June 13: The rapidly changing conditions of the past 24 hours focused the entire community on the flood. Our call center's volume jumped from 250 to 3,043 calls per day, most of them from volunteers. Volunteers and resources

FIGURE 3-3 In a river community, nearly every major road (such as Highway 6 West, also known as the Coralville Strip, shown here where it crosses Hawkins Drive and Rocky Shore Drive) intersects the floodplain at some point along its route. This results in the closure of many arterial streets and cripples remaining routes with gridlock-producing traffic. Simple tasks such as driving to work become laborious, and the delivery of emergency services is dangerously impeded. *Photograph by Richard A. Fosse, city of Iowa City.*

were focused on protecting our wells and wastewater treatment plant. The Corps once again raised the projected flood peak, to 44,000 cfs on the 16th but acknowledged the difficulty of making an accurate prediction because the Iowa River's flow upstream, at the Marengo streamgage, now exceeded all historical records. Flows coming into the reservoir thus could not be estimated with certainty.

Saturday, June 14: The weekend brought a huge turnout of volunteers to help Iowa City neighborhoods, Coralville, and the University of Iowa with sandbagging. The day was capped by another round of storms and a tornado warning. Thousands of volunteers were forced to take shelter, and when the storm passed it was too late to resume work. The most significant loss of the day was the University of Iowa's power plant. Water found its way into the utility tunnel system, which flooded the power plant from the inside and forced its closure.

Sunday, June 15: The flood peaked at 6:30 A.M. with a maximum flow through Iowa City of 41,100 cfs, slightly less than a 500-year flood. The flood peak stopped within 5 inches of closing Iowa City's two remaining bridges.

At this point, we entered a new phase of flood response characterized by vigilance. As noted earlier, being downstream from a large dam effectively puts a flood in slow motion. This is especially true for the water's recession. We knew that the first flood peak might not be the last or highest. In 1993, we experienced 10 peaks before the highest came, 35 days after the onset of the flood. We were only 12 days into the 2008 event. As a precaution, we encouraged everyone to leave sandbags in place until the water level in the Coralville Reservoir dropped well below the emergency spillway. We were fortunate. This time, the water declined slowly and steadily, with only 32 days above flood stage and no secondary peaks.

Note

1. Discharge estimates (along with other flood-related information) were received during daily phone conversations with personnel at the Corps' Coralville Reservoir office. I recorded these estimates each day and sent them to Iowa City officials for inclusion in press releases issued to keep its residents informed.

Barbara Eckstein
Rodney Lehnertz

4 The University of Iowa and the Flood

THE UNIVERSITY OF IOWA and the city of Iowa City were founded together in the 1840s. They have always shared a common main street that runs along a ridge on the east bank of the Iowa River. The university and the city have grown together, now occupying both the east bank ridge and the hills to the west and the river valley in between. Among the university buildings erected in the floodplain along the river are the main Power Plant, the Water Plant, the Main Library, the English-Philosophy Building, the Iowa Memorial Union, and the Iowa Advanced Technology Laboratories on the east side of the river; and the C. Maxwell Stanley Hydraulics Laboratory, the Art Building and Art Building West, the Museum of Art, the Theatre Building, Voxman Music Building, Clapp Recital Hall, and Hancher Auditorium on the west side of the river. These latter westside buildings comprised the Arts Campus (see figure I-4 for site locations). Mayflower Residence Hall, which houses more than 1,000 students, sits somewhat farther upriver, across from City Park. The majority of these buildings were built after the completion of the Coralville Dam, located about 8 river miles upstream from Iowa City's downtown. The dam and the reservoir it creates lie on the far side of the city of Coralville, which borders the university and Iowa City to the north.

A heritage of creativity, performance, teaching, and research is embedded in these buildings. Artist Grant Wood had a studio in the Art Building, which was built in 1936. Playwright Tennessee Williams was a student in the Theatre Building. Author Flannery O'Connor was a student in the Writers' Workshop, once located in the English-Philosophy Building. Writers Wallace Stegner, John Cheever, and Philip Roth walked the banks of the Iowa River, trying to come to grips with the culture of Iowa and Iowa City.

In the flood of 1993, the University of Iowa incurred approximately $6 million in damages (Cook 2008). University officials subsequently created a new Flood Emergency Response Plan (FERP) designed to more than meet the challenge of a similar, extraordinary event. In May 2008, following a winter of unusually heavy snowfall and then a wet spring, UI officials anticipated the need to translate the FERP into action. In June, that need became a reality.

Tuesday, June 3: UI housing staff and volunteers began building a sandbag wall around Mayflower Residence Hall because of rising waters.

Wednesday, June 4: Students staying in Mayflower were evacuated, and campus visitors were urged to avoid Dubuque Street, which runs in front of Mayflower Residence Hall, along the river.

Thursday, June 5: Because of continued heavy upstream rains, the release rate from the Coralville Reservoir was increased, and there were concerns that the reservoir might overflow the dam's emergency spillway for the first time since 1993. University staff members built sandbag walls around the Arts Campus and other threatened buildings on the east and west sides of the Iowa River.

Friday, June 6: A call went out for volunteers to help with sandbagging operations. The university initiated a seven-day-a-week administrative flood meeting to help coordinate the latest conditions and efforts to protect the campus.

Saturday, June 7: Administrators issued an alert to the Arts Campus, warning of the need to prepare for relocation of classes, staff, and materials of value.

Monday, June 9: Normal activities on the Arts Campus were suspended. Administrators in the Provost's Office and the College of Liberal Arts and Sciences, together with the space manager in Facilities Management, made plans to relocate offices of all buildings along the west bank of the Iowa River. Meanwhile, the Museum of Art implemented its disaster plan, preparing items to be transported off campus.

Tuesday, June 10: The Coralville Reservoir topped the dam's emergency spillway, and university officials evacuated the Arts Campus.

Wednesday, June 11: Water began backing up into the Arts Campus (figure 4-1). Based on the U.S. Army Corps of Engineers' (Corps') flood projections,

FIGURE 4-1 Photograph of the University of Iowa Arts Campus taken June 16, when the Iowa River stage was around 31 feet, nine feet above flood stage. The new Art Building West, completed just two years before the 2008 flood, is in the foreground right. The Museum of Art, directly to the left, normally sits along the banks of the Iowa River. Additional flooded university buildings line the far shore of the river. *Photograph by Tom Jorgensen, University of Iowa Office of University Relations Publications.*

which made clear the fate of the Arts Campus, and to address access and safety concerns, sandbagging efforts were shifted and consolidated to the east side of the river. The university made plans to evacuate and close 12 flood-threatened buildings by Friday afternoon. Late in the evening, after a plea from UI President Sally Mason, the National Guard arrived.

Thursday, June 12: The water exceeded the 1993 flood level. The Iowa Memorial Union (IMU) offices and the University Book Store, located in the IMU, prepared to evacuate, even as parents and freshmen completed an orientation session in the building. Earlier that day, a widespread storm produced significant rainfall in the Iowa River basin, and the river was rising faster than expected, with Corps flood prediction models becoming more dire as the day passed.

Friday, June 13: Volunteers formed a book brigade to move approximately 10,000 books in the Main Library basement to higher levels. UI summer session classes, camps, and events were suspended through the following week, additional floodplain buildings were evacuated, and employees not involved in patient care, utilities, security, or facilities were asked to stay home through the following week.

Saturday, June 14: The main Power Plant was shut down at 2 A.M., due to flooding from within. Water infiltrated the utility tunnels adjacent to the river and despite desperate efforts to stop the flow, the Power Plant took on more than 21 feet of standing water (figure 4-2). The tunnels, part of a 5.9-mile underground system that provides steam throughout the campus, contributed to flooding in several buildings along the river. Utilities staff began a round-the-clock effort to construct two miniature power plants, one on each side of the river, from parts delivered from around the country. Construction was completed in three days. Back-up utility systems, located high on the river's west bank at the University of Iowa Hospitals and Clinics, were activated and because flooded highways limited access to Iowa City, flights were initiated to

FIGURE 4-2 The university's main Power Plant, built on the east shore of the Iowa River, took on more than 21 feet of standing water, which entered the building through underground utility tunnels. The plant is shown here on June 16, the day after the flood crested in Iowa City. *Photograph by Tom Jorgensen, University of Iowa Office of University Relations Publications.*

bring critically needed nurses into the hospital from the Quad Cities. More than 2,000 volunteers continued to build sandbag walls on the east side of the river (figure 4-3). Students, faculty, and staff were joined by a diverse collection of community members, including Amish citizens from Kalona and prisoners from the county jail. Johnson County was declared a federal disaster area.

Sunday, June 15: Four more UI buildings were reported flooded. Despite concerns that river levels could rise another two feet, it was announced that the river had crested, and the water level began to drop slowly.

The following day, companies specializing in flood clean-up and recovery began work in and around university buildings. More than 500 tons of sandbags were collected and sent south to Iowa communities still fighting the flood. In all, 22 major buildings had flooded. The following Monday, June 23, summer session resumed with classes relocated to buildings not affected by the flood. Preparations began immediately so the fall semester could start as scheduled

FIGURE 4-3 On June 14, the day before the flood peaked, more than 2,000 students and other volunteers sandbagged university buildings on the east side of the river. *Photograph by David B. Jackson, University of Iowa.*

in August. Many classes had to be relocated, but none were canceled. The rallying cry for Facilities Management, the Provost's Office, student services, the registrar, and many others was "August 15th," the date by which the university needed to open its doors to all of its 30,000 students.

The flood of 2008 was significantly larger and more destructive than the flood 15 years earlier. One measure of that damage can be expressed in dollars. The flood of 2008 inflicted $232 million[1] in damages to university facilities and their contents (UI 2008), nearly 40 times the amount incurred in 1993. It displaced all the occupants of the floodplain buildings mentioned above, some for weeks, others for months. Some may be displaced for years. The University of Iowa and Iowa City, together with Coralville and all communities in the watershed, are now thinking through a new relationship with the Iowa River.

Note

1. $232 million is the estimate, from September 2008, of flood damages on campus. While this figure accurately represents the flood damage, in March 2009 the University of Iowa reported to the Board of Regents, State of Iowa, the total anticipated cost of the flood, including business interruption, leased replacement space, long-term protection of recovered buildings, and replacement of severely flood-damaged facilities, to be $743 million.

References Cited

Cook, Diana. 2008. Former Director of Risk Management, University of Iowa. E-mail communications to Rodney Lehnertz, December 2008.

University of Iowa (UI). 2008. Flood Recovery Report for September 9, 2008, p. 22. http://www.uiowa.edu/floodrecovery/recovery-reports/index.html.

Linda Langston

5 Linn County and the Flood

~~~~~~

A NEAT RECTANGLE in eastern Iowa, Linn County (population 200,000) is the state's second most populous county and one of its largest manufacturing centers. The Cedar River flows across the county from northwest to the southeast, cutting through its largest city, Cedar Rapids (population 125,000).

During the 157 years before 2008, the Cedar River flooded often. In 1851 and 1929, the river in Cedar Rapids rose to a stage (river height) of 20 feet—8 feet above the flood stage. It reached nearly that high in 1961 and 1993, and has exceeded a stage of 17 feet ten more times since 1851 (USGS 2008). After heavy winter snowfalls and a wet, cool spring, 2008 was expected to become another heavy flood year. Here is the story of Linn County's response to that threat.

Sunday, June 8: In anticipation of projected serious flooding, the Linn County Board of Supervisors called a meeting of all department heads and elected officials to discuss necessary precautions. Projections for the flood crest were 20 feet (NWS 2008). Rainfall was heavy.

Monday and Tuesday, June 9 and 10: The projected flood stage had already risen to 22 feet (NWS 2008). Sandbagging took place in many Linn County locations. County roads personnel directed sandbagging efforts on the Cedar

Rapids levee, which stretches almost 2 miles through and beyond the downtown area on the west side of the river and is designed to protect up to a flood crest of 22 feet. The levee was sandbagged to almost 25 feet. Throughout the county, people were observed to be calm and seemingly prepared for the potential of a 22-foot flood crest.

Within Cedar Rapids, the sandbagging efforts of volunteers and city and county staff were centralized at the Public Works building on Sixth Street SW (see figure I-3 for site locations). In the small town of Palo (population 424) about 6 miles north of Cedar Rapids, sandbagging efforts were organized at the community center in the middle of town. The Duane Arnold Nuclear Power Plant is located just north of Palo. Flooding there was not anticipated, but power plant personnel carefully monitored river levels.

Wednesday, June 11: At 2:45 A.M., the Linn County Emergency Operations Center (EOC) was totally operational. The EOC is a standing emergency center located at Kirkwood Community College. A small staff is usually present, but the EOC has the room and resources for many more during times of natural disaster or emergency. In 2008, city and county elected officials, state and local law enforcement officers, the National Guard, health department officials, hospital and education officials, and transportation, engineering, GIS (geographic information systems), Red Cross, and Salvation Army personnel all used this centralized location. The EOC ran on a 24-hour basis and at any given time 125 to 150 people were on-site, forming a beehive of activity and coordinating all emergency operations. Cedar Rapids Fire Chief Steve Havlik was the Incident Commander for the emergency and directed all operations. Within the EOC, all officials and volunteers had access to weather and river information from the U.S. National Weather Service's regional River Forecast Center, news broadcasts, and communication equipment, as well as hook-ups for wireless computers and many telephones. Through the EOC, emergency shelters for evacuees were set up in Cedar Rapids schools. Transit systems activated to evacuate people with special needs, including people with disabilities, special medical needs, and the frail elderly, from areas anticipated to flood.

By midmorning, the river's stage reached 24 feet and preparations were being made for a higher flood crest. Water began to overwhelm the sewer systems in the Time Check and Czech Village neighborhoods of Cedar Rapids, and people began to evacuate their homes (figure 5-1). Many businesses in downtown Cedar Rapids closed in early afternoon to get people home and off the streets before floodwaters rose, though few anticipated that floodwaters would reach into the downtown's central business district.

FIGURE 5-1 Water encroaches on homes in the Czech Village neighborhood in Cedar Rapids. *Photograph by Mike Duffy, Linn County Secondary Road Department.*

Thursday, June 12: Rainfall was heavy. The changing predictions of the height and time of the anticipated flood crest continued to challenge everyone. On this one day, we saw flood crest predictions rise from 24 to 26 to 28 feet (NWS 2008). Spreading floodwaters overtopped the levee on the west side of the Cedar River and began to engulf the Time Check neighborhood. Due to ever-increasing flood projections, GIS was used to plot maps of areas that would flood should the flood crest rise to 30 or 35 feet. This information assisted police and fire departments in developing evacuation plans on a block-by-block basis.

By evening the Linn County jail was evacuated. The buses that left the jail were the last vehicles on the city's downtown bridges. One bridge, the railroad bridge near 8th Avenue, collapsed unexpectedly into the river. The water continued to rise until all the downtown bridges on the Cedar River (excluding the Interstate 380 bridge) were closed, and water eventually rose over the bridge decks.

Challenges to the Cedar Rapids well system became apparent by evening. Of the 49 well fields that provide water to Cedar Rapids, only one remained operational, and water was about to inundate its pumps, as well. City and county staff called the EOC and requested a media alert for sandbagging volunteers. Over 1,000 volunteers responded, coming out in driving rain and lightning to help preserve this one last well, the only remaining water supply for the community. A water supply was also vital to operations of the Duane Arnold Nuclear Power Plant, which requires a ready supply for cooling. The plant was dropped to 20 percent of its operational capacity as a precautionary measure, in case the water supply was compromised. The Cedar River ultimately came within a quarter mile of the plant.

Volunteers also went to Mercy Hospital—almost 10 blocks from the Cedar River—in an attempt to prevent its inundation. The GIS maps developed for emergency planning purposes made it clear that the hospital would flood. Despite the sandbagging, it was evacuated in the early morning hours of June 13, after the lower levels flooded late in the evening of June 12.

Friday, June 13: In the morning the Cedar River crested in Cedar Rapids at 31.12 feet, with peak flood discharges reaching an astounding 140,000 cubic feet per second (cfs). During the four largest floods of the past 157 years, the river's peak flow rates in Cedar Rapids had been about half that, between 65,000 and 73,000 cfs (see figure 7-2). The river's 31.12-foot flood stage in Cedar Rapids was more than 11 feet above the historic high of 20 feet and 19 feet above the flood stage of 12 feet (USGS 2008). Floodwaters remained high through June 17.

Road closures and road and bridge washouts hampered transportation throughout Linn County. Interstate 380 provided the only way to move vehicles through Cedar Rapids. Interstate 380 closures south of Linn County made it necessary to designate alternate routes on gravel roads. These roads were not designed for the large trucks and buses that used them from June 12 through June 18, nor were they designed for the amount of traffic during this period. They were considerably longer and traffic moved more slowly than on the normal routes.

The Cedar River slowly retreated to its banks, leaving behind debris and devastation (figure 5-2). People who chose to stay in their homes were rescued from their rooftops. Homeowners and business people alike were frantic to find out the disposition of their properties. Keeping people safe and calming their fears at this time was a Herculean task. It required two or three news conferences daily to inform the public about what was going on and what they could and should do. In spite of being exhausted, city and county officials worked together to provide timely information and to assure residents of the continuity of services. The water system of Cedar Rapids remained tenuous

FIGURE 5-2 Houseboats swept away from Ellis Harbor piled up along the railroad bridge between First Street NW and the Quaker Oats plant in downtown Cedar Rapids. *Photograph by Mike Duffy, Linn County Secondary Road Department.*

and officials encouraged citizens to limit water use and refrain from showering or otherwise using water unnecessarily.

In the end, 25,000 people were evacuated from their homes in Linn County, including the entire town of Palo. In Cedar Rapids, water covered 1,300 city blocks or 9.2 square miles, including almost the entire downtown business district (figure 5-3). As much as 12 feet of water inundated homes and businesses. Over the next three weeks, the EOC continued to effectively protect the people and infrastructure of Linn County. Having all the decision-makers in the same room, with the same goals of keeping people safe and with ample opportunities for communication, proved invaluable to our successful mission. Also crucial were the practice emergency drills that our local communities carried out before the flood; these were required frequently because of our proximity to a nuclear power plant. Drills brought everyone together and helped people to learn their roles. Relationships that had been formed during these drills assisted in getting

FIGURE 5-3 Downtown Cedar Rapids (seen here from the Interstate 380 bridge, the only bridge to remain open) was flooded to depths of 12 feet. The inundated building in the photo's right is the Cedar Rapids City Hall. *Photograph by Jason C. McVay, U.S. Geological Survey.*

the necessary work accomplished during the flood. Practice and cooperative efforts of all county personnel helped ensure that there were no fatalities, despite the tremendous magnitude and force of the 2008 flood.

Acknowledgments

I would like to acknowledge Steve Havlik, fire chief of Cedar Rapids, who served as Incident Commander for the Flood of 2008 in Cedar Rapids. His calm and effective direction assured that no lives were lost and that all of us in the EOC were given the support we needed. I would also like to acknowledge Kirkwood Community College and Linn County, whose collaborative efforts built the state-of-the-art EOC and enabled emergency personnel to have the physical and technical assistance needed to protect the public. Thanks, also, to the Emergency Management staff, Linn County staff, and City of Cedar Rapids staff whose fine work kept people safe.

References Cited

U.S. Geological Survey (USGS). 2008. Water Resources Division, National Water Information System: Web Interface, Peak Streamflow for Iowa, Cedar River at Cedar Rapids, IA. http://nwis.waterdata.usgs.gov/ia/nwis/peak. Accessed December 2008.

U.S. National Weather Service (NWS). 2008. NOAA, North Central River Forecast Center. www.crh.noaa.gov/ncrfc/content/river_fcsts.php. Accessed continuously during the June 2008 flood.

A. Allen Bradley, Jr.

6 Forecasting a Record Flood

IN IOWA, NEARLY EVERYONE has used weather forecasts to make everyday decisions—such as how to dress for the day, or whether it is time to plant crops. We know that weather forecasts are not perfect. But from our everyday experience, we also know that forecasts do not have to be perfect to be useful.

In contrast, the public looks at river forecasts only when a river is expected to overflow its banks. When it comes to using a flood forecast to make a decision, the stakes are very high—such as how to place sandbags to keep a building dry, or whether it is time to evacuate homes and businesses.

The National Weather Service (NWS) is responsible for issuing river forecasts and flood warnings to the public (Mason and Weiger 1995). River forecasts predict the river stage (the future height of the water surface above a base elevation, such as the streambed) and the river flow or discharge (the rate at which the water is moving). Flood warnings predict the severity of an anticipated flood—minor, moderate, or major flooding. And when anticipated flood stages are higher than any recorded in historic times, the NWS will issue a warning for a record flood. During the Iowa floods of 2008, the NWS Forecast Offices in the Quad Cities and Des Moines issued hundreds of river forecasts and flood warnings for locations along the Cedar and Iowa Rivers.

From the forecaster's perspective, a record flood event is probably the most difficult river-forecasting situation imaginable. A record flood is uncharted territory; past flood experience is no longer a reliable guide as to what might happen. The forecasters are also under the intense scrutiny of emergency managers and the public, many of whom are looking at river forecasts for the first time. Everyone wants the forecasts to be perfect. Some even make decisions as if they were. But forecasting a flood that is larger than anyone has ever experienced is fraught with uncertainty.

Imagine what it would take to make a perfect flood forecast. First, you would need a perfect weather forecast to know exactly how much rain would fall and where. Then, you would need to track how much of the rainwater flowed across the landscape, and its path. Once the water entered a river, you would need to calculate how quickly it was moving and the obstacles it would encounter—levees, sandbags, bridges, even buildings—as it spread across the floodplain. Such precise accounting of water exceeds our scientific ability, making a perfect forecast a practical impossibility.

Given these inherent limitations, just how does the NWS make river forecasts? The NWS continuously monitors river and weather conditions, and combines this information with imperfect weather forecasts, to make approximate numerical predictions of future river flows and stages. The forecasters refer to these predictions as numerical guidance, because judgment and experience also play a role in issuing river stage forecasts and flood warnings.

River forecasting is a data-intensive process. Perhaps the most important data are recent observations of river stages and flow rates. These tell the forecaster how much water is already in the river and making its way downstream. For the most part, river data used in forecasting come from U.S. Geological Survey (USGS) streamgages (see plate 1), where river stage and flow rates are continuously monitored.

Precipitation observations are also crucial to river forecasting. These data are obtained from both rain gauges (cylindrical buckets designed to record rainfall accumulation) and NEXRAD weather radar estimates. These tell the forecaster how much water has recently fallen—critical information in estimating how much water will run off and soon reach the river. These data also let the forecasters check the validity of recent precipitation forecasts.

River and precipitation data are transmitted by satellite, landline, and computer network to one of the NWS's 13 River Forecast Centers. Hydrologists there assemble all available observations, combine them with weather forecasts, and feed them into computer models that simulate watershed processes and river hydraulics (Braatz et al. 1997). The computer model subdivides a river basin into smaller drainage areas known as forecast segments. Precipitation observations and forecasts are used to predict the amount of runoff. Runoff from each forecast segment enters the river, where it combines with water already in the river from upstream. The river hydraulics model then computes how quickly that water volume will move downstream through the network of river channels.

Each day the forecasters at the NWS River Forecast Centers use their computer models to make river stage forecasts at thousands of locations in the United States. First they predict river stages ahead for several days; then they watch for several hours to assess how well their forecasts match actual river stage observations. If observed and forecasted data do not match, forecasters try to diagnose why not and make corrections when making the next forecasts. So as more information comes in, the forecasters adjust their predictions to anticipate the changing river conditions. The predictions made at a River Forecast Center are shared with NWS Forecast Offices. There, hydrologists and meteorologists use the information to make the official forecasts and warnings that are issued to the public (and are available online at www.weather.gov).

By the early spring of 2008, NWS forecasters were anticipating potential flooding in eastern Iowa when they issued their *Spring Hydrologic Outlook*.[1] The cold and snowy winter was a precursor to the record flooding, and forecasters were keeping a watchful eye on the unusually large snow pack in northeastern Iowa. By late spring, most eastern Iowa streamgages were reporting unusually high river levels. With saturated soils, above-average spring rainfall, and cooler than average conditions, minor flooding had already occurred at several locations on the Cedar and Iowa Rivers.

Heavy rainfall throughout Iowa beginning on Monday, June 2, again pushed river levels toward flood stage. By Thursday, June 5, the NWS Forecast Offices in the Quad Cities and Des Moines had issued warnings for minor or major flooding along the Cedar and Iowa Rivers.[2] The situation rapidly worsened. By Saturday, June 7, floodwaters in the upstream portions of the Cedar and Iowa River watersheds were rising from the recent heavy rainfall, and with more rainfall on the way, the NWS was forecasting major flooding at Cedar Falls, Cedar Rapids, and Iowa City.

The continuing rains loomed large in the forecasting process. In making a river forecast, the NWS River Forecast Center uses a forecast of rainfall accumulations anticipated for the next twenty-four hours (precipitation forecasts further out are not reliable enough for river forecasting). Precipitation forecasts can do a good job of predicting the geographical region of future heavy rainfall (see plate 2). However, when thunderstorms develop, the heaviest rainfall is highly localized, and observed rainfall accumulations are often much greater than the forecast. For example, the rainfall forecast issued for June 9 (plate 2) called for a 24-hour precipitation accumulation between 1 and 1.5 inches (a significant amount) for the Cedar Rapids area.[3] However, rainfall totals between 2 and 4 inches were observed.[4] The tendency to underpredict future rainfall accumulations continued throughout the flood.

As the flood event unfolded, the NWS Forecast Offices held open conference calls with emergency managers and public officials. The calls provided up-to-the-minute information on observed conditions, rainfall forecasts for the upcoming days, and warnings of how the situation could worsen if actual rainfall was greater than forecast. These calls disseminated flood news quickly, enabling communities to prepare for what might happen in the days ahead.

With the continuing rains and worsening prognosis, the river crest forecasts continued to be adjusted upward (figure 6-1). By midday on Sunday, June 8, even before the Cedar River had reached flood stage, the NWS upgraded its forecast from major to record flooding at Cedar Rapids. And by midday on Monday, June 9, the NWS was forecasting record flooding for Cedar Falls and Iowa City[5] several days before the rivers would crest.

Forecast models predict how much water will be flowing in the river; however, the public is interested in river stage—how high a flooding river will rise. To predict river stage, forecasters use past river measurements to construct a rating curve, which displays an empirical relationship between river stage and flow rate (figure 6-2). These past measurements are typically made by the USGS as a routine element of its streamgage monitoring program. Rating curves have an inherent limitation: they extend out only to the largest flow rates and river stages observed in the past (figure 6-2). As time would tell, when the Cedar River finally crested in Cedar Rapids, its flow rate would be twice as large as anything previously measured, and the river stage would be almost 11.5 feet higher. With record flooding anticipated at many forecast locations, forecasters scrambled to extend existing rating curves so they could be applied to the extreme 2008 floods. At the North Central River Forecast Center, personnel from the USGS were reassigned to work side by side with the NWS during the flood and extend rating curves for locations along the flood-impacted rivers. The task proved difficult to accomplish accurately, especially at Cedar Rapids; several revisions were needed, because the USGS measurements made as the river rose showed that flows were not spreading out over the floodplain

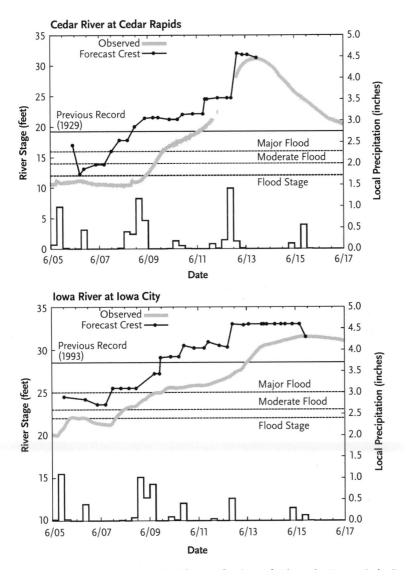

FIGURE 6-1 The evolution of the 2008 forecast flood crest for the Cedar River at Cedar Rapids and the Iowa River at Iowa City. The forecast flood crest is shown at the time when the forecast was issued; the crest is the highest river stage expected sometime over the next seven days (the forecast time horizon). Also shown is the actual observed river stage at the time that the forecast was issued. The difference indicates the anticipated rise in the river stage from the current time to whenever the crest occurs. The bars on the bottom of the graph indicate the actual six-hour precipitation accumulations in the vicinity of the forecast location. Forecasters utilized observations of past precipitation and river stage for each forecast, as well as a forecast of the precipitation amount expected 24 hours into the future. For the Cedar River, the gap in river stage operations (6/11 to 6/13) reflects the loss of the Cedar Rapids streamgage to rising floodwaters. *Illustration by the author, based on NWS data as explained in footnote 2.*

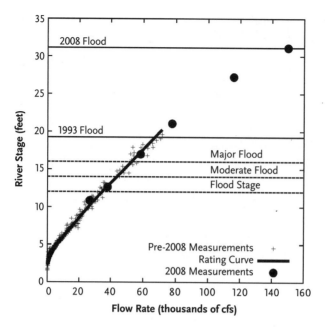

FIGURE 6-2 Rating curve, relating river stage to flow rate for the Cedar River at Cedar Rapids. Direct river measurements made by the U.S. Geological Survey (USGS) of river stage and flow rate at the streamgage site (shown as crosses). A rating curve (shown as a solid line) is then established over the range of river observations. USGS direct measurements made in 2008 (shown as black circles), including several taken during the flood event itself, illustrate how the river's stage and flow rate exceeded the previously defined rating curve and pushed forecasters into uncharted territory. *Illustration by the author, based on USGS field measurements available at water.usgs.gov/waterwatch/.*

as much as first anticipated. Another unanticipated challenge was the temporary loss of river observations at the Cedar Rapids and Vinton streamgages. On Thursday, June 11, the USGS streamgage at Cedar Rapids lost power and later was completely inundated. Upstream the Benton County streamgage at Vinton was raised above floodwaters by the hasty addition of a marked pole, but later it too was submerged. The forecasters were now operating in the dark, having lost their best information sources on actual river conditions. And the situation only got worse.

By the morning of Friday, June 12, heavy rains were again falling, adding more runoff to an already rising river in Cedar Rapids. At about the same time, the Railroad Bridge near 8th Avenue collapsed into the river—something forecasters could not anticipate and then only could speculate as to whether it might have an effect. Through the extraordinary efforts of the USGS, a temporary streamgage was reestablished in Cedar Rapids and began transmitting data about 20 hours before the river crested. The restoration of river observations enabled the NWS to revise its river forecasts upward to the unimaginable levels that finally occurred. But the warning came as a surprise to many and too late for some to react.

How useful were the forecasts for this record flood event? In hindsight, it is impressive that forecasting technologies were capable of predicting an unprecedented record flood with several days advance warning. On the other hand, for those facing agonizing decisions about how to protect life and property, the continuously adjusted forecasts were a source of frustration, even though that is the nature of river forecasting.

How could forecasts be improved and made more useful in the future? We can establish better ways to convey forecast information and to communicate the inherent uncertainty of river forecasts. Some ideas along these lines are already in the works (McEnery et al. 2005; NWS 2002), such as providing the public with online access to flood inundation maps for specific forecast conditions, and issuing forecasts that include a range of likely outcomes.

Forecasters will state that they need more observations to make better forecasts. Yet across the United States, the streamgage network used for forecasting is actually declining (NRC 2004), so forecasters are being asked to do more with less information. The true costs of eliminating monitoring stations and capabilities will become apparent when we face disasters for which better forecasts could have minimized losses and destruction.

Despite all these problems, forecast technology has evolved significantly in the 15 years since the 1993 floods. Forecasts of river flows and stages have been extended from three days to seven days into the future. Forecasts of precipitation were not available for use in river forecasting in 1993, nor were precipitation data from weather radars. Even though future enhancements of forecasting technologies will never make river forecasts perfect, there is hope that additional technological advances will help emergency managers and the public make better decisions when the next record flood occurs.

Notes

1. The *Spring Hydrologic Outlook* product describes the spring flooding expected due to snowmelt. The outlook is issued by the North Central River Forecast Center in February and March and is available online at http://www.crh.noaa.gov/ncrfc/.

2. Information on river forecasts and warnings was obtained from *Flood Warning*

and *Flood Statement* products issued in May–June 2008 from the NWS Forecast Offices in the Quad Cities and Des Moines. Recent warnings and statements are available online at www.weather.gov, and through other data transmission services for emergency managers and experts in weather and hydrology. They are archived at the National Climatic Data Center (NCDC).

3. The precipitation forecasts for the North Central River Forecast Center were obtained from the NWS Hydrometeorological Prediction Center (HPC) National Precipitation Verification Unit, available online at www.hpc.ncep.noaa.gov/npvu/.

4. The precipitation observations are North Central River Forecast Center products created for river forecasting using rain gauge and weather radar observations, available online at http://www.crh.noaa.gov/ncrfc/.

5. At the time of the flood, the NWS record flood stage at Iowa City was defined by the 1993 flood. This event was the largest observed in Iowa City since Coralville Reservoir was completed in 1958 and the reservoir's operations began to reduce downstream flooding, essentially resetting the flood record. Before 1958, when the Iowa River was not regulated by a flood control reservoir, Iowa City experienced two floods larger than that observed in 2008.

References Cited

Braatz, D. T., J. B. Halquist, R. J. Warvin, J. Ingram, J. J. Feldt, and M. S. Longnecker. 1997. "NWS hydrologic products and services: Moving from the traditional to the technically advanced." In *Preprints, 13th International Conference on Interactive Information and Processing Systems for Meteorology, Oceanography, and Hydrology.* Long Beach, Calif.: American Meteorological Society, J63–J69.

Mason, R. R., Jr., and B. A. Weiger. 1995. "Stream gaging and flood forecasting: A partnership of the U.S. Geological Survey and the National Weather Service." Reston, Va.: U.S. Geological Survey Fact Sheet FS 209-95. http://water.usgs.gov/wid/FS_209-95/mason-weiger.html.

McEnery, J., J. Ingram, Q. Duan, T. Adams, and L. Anderson, 2005. "NOAA's Advanced Hydrologic Prediction Service: Building pathways for better science in water forecasting." In *Bulletin of the American Meteorological Society* 86(3), 375–385.

National Research Council (NRC). 2004. *Assessing the National Streamflow Information Program.* Washington, D.C.: National Academy Press.

National Weather Service (NWS). 2002. *Advanced Hydrologic Prediction Services: Water Prediction for Life Decisions.* Silver Spring, Md.: NOAA/PA 200258. www.weather.gov/os/water/ahps/pdfs/NOAANWS_Brochure.pdf.

David Eash

7 Estimating Flood Frequency

SOON AFTER THE 2008 floods in Iowa reached their peak, the media labeled them as 100-year floods or 500-year floods. What exactly do these terms mean, and how do they apply to Iowa's floods of 2008?

These terms are expressions of the extreme magnitude of the 2008 floods and the frequency with which they are expected to occur. Flood frequency is expressed in several ways. One is as a probability, expressed as a percentage, that a flood of a specific level will be equaled or exceeded in any given year. Thus, a 1 percent chance flood has a 1 in 100 chance of occurring in any particular year, and a 0.2 percent chance flood has a 1 in 500 chance of occurring any year.

Another common way to express flood frequency is as the time between floods of a certain magnitude (known as the recurrence interval). If a specific flood level has a 1 in 100 chance of occurring in a given year (a 1 percent chance flood), then the average recurrence interval between floods of this size is 100 years. This flood level is called the 100-year flood. Likewise, if a specific flood level has a 1 in 500 chance of occurring in a given year (a 0.2 percent chance flood), the average recurrence interval between floods is 500 years, and it is called a 500-year flood. By this definition, a 500-year flood is larger than a 100-

year flood, which is larger than a 50-year flood, and so on. Larger floods occur less frequently than smaller floods, have a smaller chance of occurring in any particular year, and have a longer average recurrence interval.

The terms "100-year flood" and "500-year flood" are often misunderstood: people tend to believe this level of flood should only occur once every 100 or 500 years (Dinicola 1997). However, the 100-year terminology does not imply that there is a regular or uniform time interval between flood events or that a given 100-year period will contain one and only one 100-year flood event (Subcommittee on Hydrology 2008). The truth is that there is a chance that an unusually large flood will occur any year—it can occur next year, or the year after, or even in two sequential years.[1]

Flood-frequency estimates for streamgage locations are based on measurements of a river's annual maximum discharge—that is, the peak rate of streamflow that moves past a point in the river during the course of a one-year period. Discharge is expressed in cubic feet per second (cfs); one cfs is equal to 7.48 gallons of water per second. A record of annual maximum discharges is a collection of the major flood peaks measured over a historical period. But not all annual maximum discharges are floods, as a river may not overflow its banks every year.

For most streamgage locations with a sufficient record of annual maximum discharges (at least 10 years), federal agencies estimate flood-frequency discharges using the statistical techniques outlined in *Bulletin 17B* (Interagency Advisory Committee on Water Data 1982). As additional years of data are collected at streamgages, *Bulletin 17B* estimates of flood-frequency discharges are updated and become more statistically reliable. For watersheds significantly affected by urbanization or river locations downstream from major flood-control structures—like the Coralville Reservoir and Dam, which regulates flows by storing floodwaters and controlling releases to reduce flood discharges downstream—more complex statistical methods are employed to estimate flood-frequency discharges. For ungaged river locations, regional equations are used to estimate flood-frequency discharges by relating *Bulletin 17B* flood-frequency estimates for streamgage sites to ungaged sites with similar watershed characteristics (Eash 2001). Thus, flood-frequency discharges can be estimated for any river location.

What does flood-frequency estimation tell us about Iowa's 2008 floods? Were they 100- or 500-year floods? The answer depends on the particular river location being considered. Rainfall and snowmelt, and resulting tributary streamflow, vary from one location to another, and so the level of flooding can vary significantly from one river to another and from one location to another along the same river. Consider, for example, the 2008 flood discharges and recurrence intervals for streamgages on the Iowa and Cedar Rivers listed in table 7-1 (see plate 1 for location of streamgages). Recurrence intervals listed in table 7-1 for the 2008 flood are reported as a range, except for those exceeding the 500-year flood, to indicate the considerable amount of uncertainty in these estimates.

The Cedar River had extreme floods in 2008 exceeding the 500-year flood discharge at Cedar Rapids (figure 7-1) and Waverly. In contrast, at other sites along the Cedar River, recurrence intervals range from 100 to 500 years. The 2008 flood is the largest flood on record at all Cedar River streamgages listed in table 7-1.

The Iowa River had intense floods during 2008 exceeding the 500-year flood discharge at the streamgage below the Coralville Dam and below the confluence of the Iowa and Cedar Rivers at Wapello. In Iowa City, the recurrence interval of the 2008 flood is estimated to range between 100 and 500 years. The 2008 flood in Iowa City is not the greatest on record. The Coralville Dam helped reduce the 2008 flood discharge from 51,000 cfs recorded upstream from the dam at the Marengo streamgage to 41,100 cfs downstream from the dam at Iowa City. At other locations along the Iowa River, recurrence intervals for the 2008 flood range from 25 to 500 years.

Both the Iowa City and Cedar Rapids floods were extraordinary flood events; the flood in Iowa City was a 100- to 500-year flood, and in Cedar Rapids it was larger than a 500-year flood. For the Cedar River at the Cedar Rapids streamgage, the 500-year flood discharge is estimated to be 122,000 cfs; however, the 2008 maximum discharge was 140,000 cfs, or about 15 percent greater. The 2008 maximum discharge of 41,100 cfs for the Iowa River at the Iowa City streamgage falls between the 100-year flood discharge estimate of 29,000 cfs and the 500-year flood estimate of 45,000 cfs.[2]

More than 100 years of continuous streamflow records have been collected at the Iowa City and Cedar Rapids streamgages. Figures 7-2 and 7-3 display these records and also significant flood peaks prior to the start of continuous record collection. Figure 7-2 shows that the 2008 Cedar Rapids flood was almost twice as large as the 1961 flood, which was the second-largest flood of the 158-year Cedar Rapids flood record. The third largest flood was in 1993. Figure 7-3 shows that in the past 158 years, Iowa City has had three floods that exceeded the 2008 flood—in 1851, 1881, and 1918. However, all three of these floods occurred before the Coralville Reservoir and Dam were completed in

TABLE 7-1 Summary of 2008 Flood Discharges, Recurrence Intervals, and Maximum Flood Discharges from the Annual Maximum Discharge Record for Streamgages Located on the Iowa and Cedar Rivers

Streamgage Name (and number)	Drainage Area (miles^2)	Average Discharge[a] (cfs)	Date	2008 FLOOD Maximum Discharge[b] (cfs)	Recurrence-interval Estimate (years)[c]	MAXIMUM FLOOD ON RECORD Year	Maximum Discharge (cfs)	Annual Maximum Discharge Record
Iowa R. near Rowan (05449500)	429	245	6/09	7,890	50–100	1954	8,460	1941–76, 1978–2008
Iowa R. at Marshalltown (05451500)	1,532	891	6/13	22,400	25–50	1918	42,000	1903, 1915–27, 1929–30, 1933–2008
Iowa R. at Marengo (05453100)	2,794	1,929	6/12	51,000	100–500	2008	51,000	1957–2008
Iowa R. below Coralville Dam (05453520)[d]	3,115	2,393	6/15	39,900	>500[e]	2008	39,900	1993–2008
Iowa R. at Iowa City (05454500)[d]	3,271	2,300[f]	6/15	41,100	100–500[e]	1851	70,000	1851, 1881, 1903–2008
Iowa R. near Lone Tree (05455700)[d]	4,293	3,077[f]	6/15	53,700	50–100[e]	1993	57,100	1957–2008
Cedar R. at Charles City (05457700)	1,054	771[g]	6/09	34,600	100–500	2008	34,600	1946–53, 1961–62, 1965–2008
Cedar R. at Waverly (05458300)	1,547	1,107	6/10	52,600	>500[h]	2008	52,600	2001–2008
Cedar R. at Janesville (05458500)	1,661	969	6/10	53,400	100–500	2008	53,400	1905–06, 1915–21, 1923–27, 1933–42, 1945–2008
Cedar R. at Waterloo (05464000)	5,146	3,361	6/11	112,000	100–500[i]	2008	112,000	1929, 1933, 1941–2008
Cedar R. at Cedar Rapids (05464500)	6,510	3,807	6/13	140,000	>500[i]	2008	140,000	1851, 1903–2008
Cedar R. near Conesville (05465000)	7,787	5,193	6/15	127,000	100–500[i]	2008	127,000	1940–2008
Iowa R. at Wapello (05465500)[d]	12,500	9,183[f]	6/14	188,000	>500[j]	2008	188,000	1903–2008

a. Unless otherwise noted, average discharge calculated for all years of continuous streamflow record through 2007. Some years of the annual maximum discharge record were not included in the calculation of average discharge; see *Water-Resources Data for the United States, Water Year 2007* (U.S. Geological Survey, 2008) for streamflow record used to calculate average discharge.

b. See *Water-Resources Data for the United States, Water Year 2008* (U.S. Geological Survey, 2009) for 2008 flood data.

c. Recurrence intervals determined from flood-frequency discharge estimates computed for streamgages and are reported as a range of years, or as greater than 500 years, to indicate the uncertainty of the estimates. Unless otherwise noted, recurrence intervals were determined from weighted-average flood-frequency estimates computed using both *Bulletin 17B* analyses and regional equations.

d. Streamflow in the Iowa River, downstream from Coralville Reservoir, has been regulated by the Coralville Dam since September 17, 1958. This streamgage measures regulated streamflow.

e. Recurrence intervals determined from flood-frequency discharges listed in the *Flood Insurance Study for Johnson County, Iowa, and Incorporated Areas* (FEMA, 2007). The USGS uses U.S. Army Corps of Engineers (Corps) computations for flood-frequency discharge estimates at sites where a river is regulated. For this book, the Corps recommends the use of flood-frequency discharges published in the latest FEMA flood insurance study for regulated streamgages on the Iowa River.

f. Streamflow years 1959–2007 used to calculate average discharge.

g. Streamflow years 1965–2007 used to calculate average discharge.

h. Recurrence intervals determined from flood-frequency discharge estimates calculated from regional equations only (Eash, 2001).

i. Recurrence intervals determined from flood-frequency discharge estimates computed from *Bulletin 17B* analyses only (Interagency Advisory Committee on Water Data, 1982).

j. Recurrence intervals determined from flood-frequency discharges listed in the *Flood Insurance Study for Louisa County, Iowa, and Incorporated Areas* (FEMA, 1991).

1958. As seen in figure 7-3, the reservoir reduces annual maximum discharges in Iowa City. Current estimates of flood-frequency discharges for Iowa City account for the flood control effect of the Coralville Reservoir, so the 100-year and 500-year flood discharges are now much lower than they were before the reservoir was built.

One source of confusion for many is that both the 1993 floods in the Upper Mississippi River Basin and the 2008 floods in Iowa have been described as

FIGURE 7-1 Our U.S. Geological Survey (USGS) crew measured an incredible discharge of 149,500 cubic feet per second (cfs) downstream from Cedar Rapids during the crest of the Cedar River flood on June 13, 2008. Flood discharge is always measured in person by hydrographers since automated, on-site streamgage equipment only measures the river's stage (or surface-water elevation). Because the Cedar River was about 1.6 miles wide near the streamgage in Cedar Rapids (where discharge measurements are usually made), and the river there was filled with urban obstructions and other hazards, the discharge measurement for the flood was instead made at this location about 10 miles downstream at U.S. Highway 30. This discharge was later adjusted to obtain the maximum discharge of 140,000 cfs at the Cedar Rapids streamgage. The photograph shows measurement of road overflow east of the river channel. *Photograph by Scott Strader, USGS.*

100-year (or greater) events. The conclusion is that Iowa has seen two 100-year floods in the span of only 16 years. While this statement is true for a few river locations within the state, it is not true at most river locations. For Cedar Rapids, the 1993 flood peak of 71,000 cfs was only a 25- to 50-year flood (Eash 1997). In contrast, for Iowa City, the 1993 flood peak of 28,200 cfs was indeed originally estimated to have been about a 100- to 500-year flood (Schaap and Harvey 1995), but is now estimated to be about a 50- to 100-year flood (FEMA 2007) based on a longer period of record. This change in recurrence interval for the 1993 flood at Iowa City demonstrates the uncertainty in the estimates and how change occurs as the annual maximum discharge record becomes longer.

Knowledge of the size and probability of floods is essential for managing floodplains and for designing bridges, dams, levees, and other floodplain structures. Knowledge of probable flood sizes allows engineers and planners to standardize risk factors. Estimates for 100- and 500-year floods are commonly

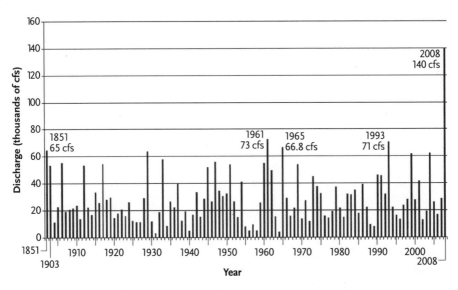

FIGURE 7-2 This graph shows the annual maximum discharge record for the Cedar River at the Cedar Rapids streamgage site, for the years 1851 and 1903–2008. Flood stage for this site in 2008 was 12 feet, which was about 33,100 cubic feet per second (cfs). The 100-year and 500-year flood estimates for this site are, respectively, 94,100 cfs and 122,000 cfs since 2008. *Illustration by the author: Annual maximum discharge data since 1903 from the streamgage, USGS, http:// nwis.waterdata.usgs.gov/ia/nwis/peak, with 1851 estimate collected on-site by local sources; 100- and 500-year estimates based on 1903–2008 data; flood stage as determined by National Weather Service Advanced Hydrologic Prediction Service, http://www.weather.gov/oh/ahps/.*

used to protect floodplain structures from erosion of the riverbed. In addition, the National Flood Insurance Program, administered by the Federal Emergency Management Agency (FEMA), manages development on floodplains using estimates for 100- and 500-year floods (FEMA 2002).

Flood-frequency estimation cannot predict when future flooding will occur. Flood occurrence is mostly an unpredictable and random process (Subcommittee on Hydrology 2008). Big floods happen irregularly because of natural climatic variations. Big floods sometimes occur in successive, or nearly successive, unusually wet years (Dinicola 1997).

The accuracy and reliability of *Bulletin 17B* flood-frequency estimation is dependent on several assumptions (Interagency Advisory Committee on Water Data 1982), most notably that past flood discharges are accurate and reliable

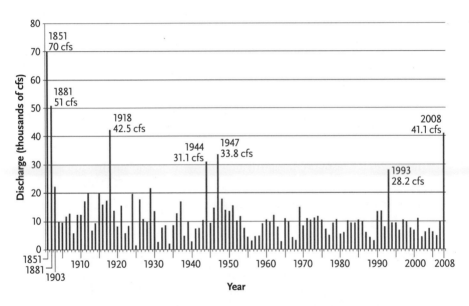

FIGURE 7-3 This graph shows the annual maximum discharge record for the Iowa River at the Iowa City streamgage site for the years 1851, 1881, and 1903–2008. Flood stage for this site in 2008 was 22 feet, which was about 13,800 cubic feet per second (cfs). The 100-year and 500-year flood estimates for this site are, respectively, 29,000 cfs and 45,000 cfs prior to 2008. Note the effect of the Coralville Reservoir and Dam in reducing annual maximum discharges in Iowa City since completion of the dam in 1958. *Illustration by the author. Annual maximum discharge data since 1903 from the streamgage, USGS, http://nwis.waterdata.usgs.gov/ia/nwis/ peak, with 1851 and 1881 estimates collected on-site by local sources; 100- and 500-year estimates, FEMA 2007; flood stage as determined by National Weather Service Advanced Hydrologic Prediction Service, http://www.weather.gov/oh/ahps/.*

indicators of future flood discharges. However, past flood discharges may not reflect future discharges if the landscape is significantly altered, or if the climate changes. *Bulletin 17B* flood-frequency estimation also assumes that the flood record is an accurate depiction of previous floods, and that the flood process can be represented mathematically as a sequence of independent annual maximum discharges that are randomly sampled from a population of all possible flood discharges. For these reasons, *Bulletin 17B* flood-frequency estimates carry substantial uncertainties (Subcommittee on Hydrology 2008).

Another significant source of uncertainty is the short historical record of large floods. At many streamgages in Iowa, the record of annual maximum discharges seems long (50 to 100 years). When it comes to a small flood, with a recurrence interval of 2 to 10 years, such a record can provide accurate estimates. But because large floods are so rare, it is difficult to accurately estimate the discharge for a flood that will occur only about once every 100 or 500 years. Hence, existing flood-frequency estimates depend greatly on the floods that have been observed in the past. Indeed, as happened following the 1993 flood (Eash 1997), estimates have increased at most streamgages that had large flood discharges in 2008. Continued operation of streamgages into the future is essential to make flood-frequency estimates more accurate and reliable.

Despite uncertainties, flood-frequency estimation remains the best tool at present for the design of structures and for the regulation of floodplains, particularly from an economic perspective. Sound flood-frequency estimation minimizes the costly overdesign or underdesign of floodplain structures. Flood-frequency estimation also remains imperative for dealing with the important problems of flood insurance and flood zoning.

Acknowledgments

The author gratefully acknowledges the technical and editorial contributions of the following individuals, whose assistance greatly improved this chapter: A. Allen Bradley, Jr., IIHR-Hydroscience & Engineering; Shirley Johnson, U.S. Army Corps of Engineers; and Edward Fischer and Robert Holmes, USGS. The information herein is based on data collected by the U.S. Army Corps of Engineers, the National Weather Service, the USGS, and several state and local agencies. Appreciation is expressed to personnel in these agencies involved with data collection, often during adverse conditions, whose efforts made this chapter possible.

Notes

1. Because the terms "100-year flood" and "500-year flood" are often misunderstood, the Federal Emergency Management Agency (FEMA) now prefers to call a 100-year flood a "1-percent-annual-chance flood" and a 500-year flood a "0.2-percent-annual-chance flood" (FEMA 2002).

2. The USGS uses U.S. Army Corps of Engineers (Corps) computations for flood-frequency discharge estimates at sites where a river is regulated. This applies to streamgages on the Iowa River below the Coralville Reservoir. For this book, the Corps recommends the use of flood-frequency discharge estimates published in the latest FEMA Flood Insurance Study (FIS) for regulated streamgages on the Iowa River. Flood-frequency discharge estimates from the current FIS for Johnson County (FEMA 2007) were used to estimate the recurrence interval for the 2008 flood at the Iowa City streamgage; discharge estimates computed for the Johnson County FIS do not include the 2008 flood discharge.

References Cited

Dinicola, K. 1997. "The '100-year flood'." Tacoma, Wash.: U.S. Geological Survey Fact Sheet FS-229-96. http://pubs.usgs.gov/fs/FS-229-96.

Eash, D.A. 1997. "Effects of the 1993 flood on the determination of flood magnitude and frequency in Iowa." Reston, Va.: U.S. Geological Survey Circular 1120-K. http://pubs.er.usgs.gov/usgspubs/cir/cir1120K.

———. 2001. *Techniques for Estimating Flood-Frequency Discharges for Streams in Iowa.* Iowa City, Iowa: U.S. Geological Survey Water-Resources Investigations Report 00-4233. http://pubs.er.usgs.gov/usgspubs/wri/wri004233.

Federal Emergency Management Agency (FEMA). 1991. "Flood Insurance Study for Louisa County, Iowa, and incorporated areas." Washington, D.C.: Department of Homeland Security.

———. 2002. "National Flood Insurance Program, program description." Federal Insurance and Mitigation Administration. www.fema.gov/library/viewRecord.do?id=1480.

———. 2007. *Flood Insurance Study for Johnson County, Iowa, and Incorporated Areas.* Washington, D.C.: Department of Homeland Security.

Interagency Advisory Committee on Water Data. 1982. "Guidelines for determining flood flow frequency, Bulletin 17B." Reston, Va.: Hydrology Subcommittee. http://water.usgs.gov/osw/bulletin17b/bulletin_17B.html.

Schaap, B. D., and C. A. Harvey. 1995. "Delineation of flooding within the Upper Mississippi River Basin, 1993—Flood of June 29–September 18, 1993, in Iowa City and vicinity, Iowa." Iowa City, Iowa: U.S. Geological Survey Hydrologic Investigations Atlas 735-B. http://pubs.er.usgs.gov/usgspubs/ha/ha735B.

Subcommittee on Hydrology. 2008. "Bulletin 17-B guidelines for determining flood frequency, frequently asked questions." Water Information Coordination

Program, Advisory Committee on Water Information, Hydrologic Frequency Analysis Work Group. http://acwi.gov/hydrology/Frequency/B17bFAQ.html.

U.S. Geological Survey (USGS). 2008. "Water-resources data for the United States, Water Year 2007." U.S. Geological Survey Water-Data Report WDR-US-2007, Iowa sites. http://wdr.water.usgs.gov/wy2007/search.jsp.

———. 2009. "Water-resources data for the United States, Water Year 2008." U.S. Geological Survey Water-Data Report WDR-US-2008, Iowa sites. http://wdr.water.usgs.gov/wy2008/search.jsp.

Why Here, Why Now?

ALTHOUGH FLOODS ARE natural processes, humanity has not entirely come to terms with this idea. Instead, since the earliest written records, people have tried to explain flooding and extreme natural events in other ways. Flood myths of worldwide deluges, often with common themes, appear widely in early writings from around the world. Consider Noah and the Ark, a story that deals with flooding as a way of cleansing iniquity from the face of the earth, after which God renews a relationship with humanity; it is recorded near the very beginning of the Bible in the sixth chapter of Genesis. The Babylonian epic of Gilgamesh, among the earliest known works of literary fiction, includes a similar story: Gilgamesh is warned of the gods' plans to destroy all life by flooding, but he survives by building a floating vessel. And fragmentary Sumarian clay tablets record a story about the gods' decision to destroy humankind in a flood; the hero Atrahasis builds a large boat, makes appropriate sacrifices, and survives to repopulate the earth.

Today we seek more concrete explanations for flooding. We ask logical questions such as what set of weather patterns produced a confluence of floodwaters.

We question whether we have modified the earth's surface and atmosphere sufficiently to create imbalances that are now bringing us watery devastation. We forget the role of floodplains and how a river's boundaries may flex. Perhaps, just as in ancient times, our modern flood stories are working to lead us back to a new relationship with the earth. Just as Noah, Gilgamesh, and Atrahasis must have done, we wonder whether we are being called to renew our balance with the earth and the creation.

Section II considers the causes of the 2008 floods and of flooding more generally. Acknowledging that rivers do flood and that floods will continue, the section looks at what we may be doing to exacerbate river flows, to increase their volume and extent. The section begins with Michael Burkart's consideration of the tremendous changes we have wrought in the land, how Iowa's nineteenth-century transformation to an agricultural landscape produced one of the most intensively managed landscapes on earth. Prior to the 1830s, Iowa was dominated by tallgrass prairies, which acted like gigantic sponges that soaked up, transpired, and stored tremendous quantities of moisture. Surface waters on this landscape would have risen or dropped slowly, and floods would have been very different from those of today. Burkart looks at the hydrology of the current agricultural landscape, which is managed to shed water as rapidly as possible and is dominated by annual row crops that cover the soils and help manage water only a few months of the year. The working landscape's lost water storage, artificial drainage systems, engineered increase in drainage channels, and resulting accelerated velocity and flow of water from the land have created a threat of stream flooding that is magnified in size and extent. In chapter 9, Wayne Petersen explains that water's rapid runoff also has been exacerbated in urban areas, where stream levels typically rise and fall rapidly. This is because cityscapes have been covered with impermeable surfaces (roads, sidewalks, roofs, etc.), and their soils have been compacted so that they absorb rain poorly. However, Petersen concludes that while urban settings produce significant flash floods in smaller local streams, their contribution to the larger 2008 floods was probably minimal simply because of Iowa's small proportion of urban land.

Many have wondered why the Coralville Reservoir did not better control the flooding in Iowa City and Coralville. John Castle describes the Coralville Dam and gives a detailed summary of the Corps of Engineers' operation of the reservoir in chapter 10. This operation is firmly set by a regulation schedule that specifies reservoir elevations and water release rates throughout the year. These regulations and the reservoir help control downstream flooding but cannot eliminate it. This fact was amply demonstrated in 2008, with its overwhelming quantities of water. Yet even then, the reservoir substantially decreased peak flooding downstream: Castle states that the maximum inflow into the reservoir was estimated at 57,000 cubic feet per second (cfs),[1] while the maximum outflow below the dam was over 17,000 cfs less or 39,900 cfs. These numbers take on significance when compared to the peak flow through Iowa City (41,100 cfs; see chapter 7), a number that could have risen to nearly 60,000 cfs if the reservoir had not existed. In chapter 11, Robert F. Sayre views the history and management of the reservoir more critically. Did the reservoir's construction create a false sense of security that encouraged floodplain development and thus increased later flood damage downstream? Sayre argues that we need a revision of the reservoir's water control plans. In particular, he argues for drawing down reservoir levels earlier in the spring, a process that would require purchase of downstream floodplain land that would be regularly flooded but could ultimately mitigate very large floods.

Eugene S. Takle tackles the broader question of climate change in chapter 12. While acknowledging that our climate is in slow but constant flux, he questions whether human influences on the atmosphere could have magnified the 2008 floods. No one can tie the floods specifically to climate change, Takle explains, but indeed the 2008 floods were consistent with modern climate trends (increasing early summer precipitation, increasing intensity of extreme weather events) and with climate models. These trends are predicted to continue. One result may be increases in early summer streamflow such as we had in 2008, with a higher probability of extreme flood events. These increases could raise the bar for current flood mitigation efforts.

In seeking causation of the 2008 floods, we find both comfort and challenge. The floods were indeed natural. There was probably little we could have done to guard against the year's extreme weather events, and flooding would have occurred regardless. Large floods that overwhelmed the prairie landscape undoubtedly occurred even before Iowa was settled. But a healthier interaction with the modern landscape and different floodplain policies could have lessened the floods' extremes and damages. And with our continued burning of fossil fuels and release of greenhouse gasses, we may be enhancing imbalances in the water cycle that will evoke still more such extreme events in coming years. This knowledge gives ample reason for considering the remediation activities outlined later in the book. ≋

Note

1. Note that the text of chapter 7 states that the maximum reservoir inflow was 51,000 cfs, a figure derived at the Marengo streamgage upstream from the reservoir. Castle's inflow at the reservoir is higher because it includes estimates of inflows from streams between Marengo and the reservoir.

Cedar River
at Charles City

Cedar River
at Waverly

Cedar River
at Janesville

**Cedar River
Watershed**

Cedar River
at Waterloo

Cedar River
at Cedar Rapids

Cedar Falls /
Waterloo

Iowa River
near Rowan

Linn

Iowa River below
Coralville Dam

Iowa River at
Marshalltown

Cedar
Rapids

**Iowa River
Watershed**

Coralville

Iowa River
at Marengo

Iowa
City

Johnson

Cedar River
near Conesville

Iowa River
at Iowa City

Iowa River near
Lone Tree

Iowa River
at Wapello

**Iowa–Cedar
Watershed below
Confluence**

0   50   𝒩

miles   ⋏

PLATE 1 The river networks and major cities of the Iowa and Cedar River watersheds. This map shows the 13 U.S. Geological Survey (USGS) streamgage sites on these rivers in Iowa. Additional streamgages not shown on this map are located on the rivers' tributaries and in Minnesota. Streamgages serve as our eyes on the rivers, monitoring the water's stage (level) and discharge (flow) and creating a data base for the river at that site that feeds into flood forecasts. This plate is discussed in the Introduction to Section I, and data from these streamgages are included in the several chapters. Location of the watersheds is shown in figure I-2. *Illustration by Casey Kohrt, Iowa Geological and Water Survey, based on USGS stream-gages and watershed boundary data sets.*

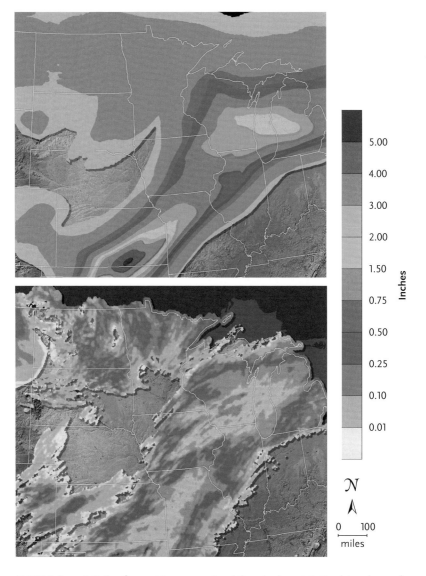

PLATE 2 A precipitation forecast (upper map) and the actual precipitation later observed (lower map) are both vital information used in forecasting river levels during a flood. This upper Midwest example shows the precipitation forecast and its corresponding observation for the 24-hour period ending at 12 UTC (or 7 A.M. CDT) on June 9, 2008. The precipitation forecast was produced at the NWS North Central River Forecast Center and was used by the river forecasters to predict future river stages at hundreds of locations in the upper Midwest. In this case, the areas where rain was forecast, and the regions where heavy rainfall was expected, generally match what was actually observed (lower map) with the NEXRAD weather radar and rain gauge measurements. However, at the scale of an individual river basin, the difference between the forecast and observed rainfall can be large. This leads to errors in the timing and severity of forecast flooding. This plate is discussed in chapter 6. *Illustration by Radoslaw Goska, adapted from data products used operationally by the NWS.*

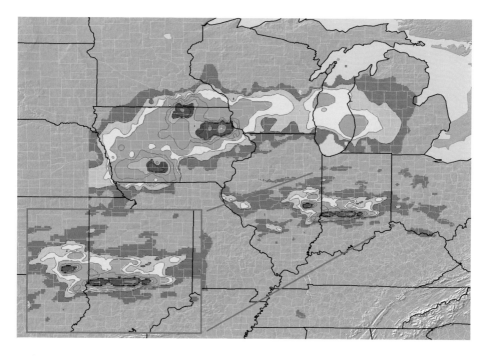

5/23/2008 (8 a.m. EDT) to 6/12/2008 (8 a.m. EDT)

☐ >1/2	☐ 1/10–1/20	☐ 1/100–1/150	☐ 1/500–1/1000
☐ 1/2–1/5	☐ 1/20–1/50	☐ 1/150–1/200	☐ <1/1000
☐ 1/5–1/10	☐ 1/50–1/100	☐ 1/200–1/500	

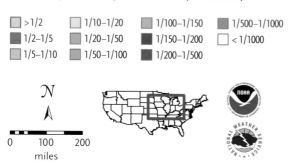

𝒩
↑
N

| | | |
0   100   200
miles

PLATE 3 Annual exceedance probability for 20-day rainfall totals. Note that some regions of eastern Iowa experienced conditions that statistically would occur only once in 200–500 years (purple) or once in 500–1,000 years (pink). Isolated points had 20-day rainfall totals that statistically would occur less than once in 1,000 years (white). This figure is discussed in chapter 12. *Illustration provided courtesy of Geoffrey Bonnin, NOAA 2008.*

Estimated Rainfall on 12 April 2008 – 12 June 2008 (sum)

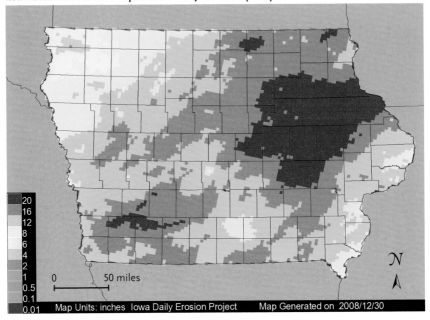

PLATE 4 *Above:* Estimated rainfall exceeded 20 inches in some eastern Iowa areas during the month leading up to the floods of 2008. This figure is discussed in chapter 16. *Illustration from IEM 2009, based on Cruse et al. 2006.*

PLATE 5 *Opposite top:* Average estimated runoff for townships in Iowa from 12 April–12 June 2008 was very high, leading to the intense flooding observed in eastern Iowa. This figure is discussed in chapter 16. *Illustration from IEM 2009, based on Cruse et al. 2006.*

PLATE 6 *Opposite bottom:* Average estimated soil loss for townships in Iowa for June 12, 2008. Erosion rates were very high for this single day, and in some townships they exceeded the tolerable amount of erosion for a complete year. This figure is discussed in chapter 16. *Illustration from IEM 2009, based on Cruse et al. 2006.*

Average Runoff: 12 April 2008 – 12 June 2008 (sum)

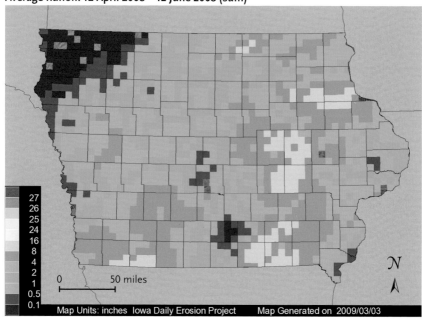

Average Soil Loss: 12 June 2008

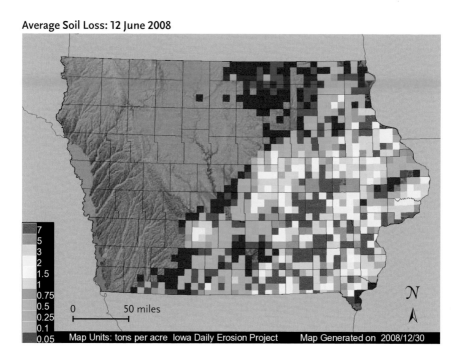

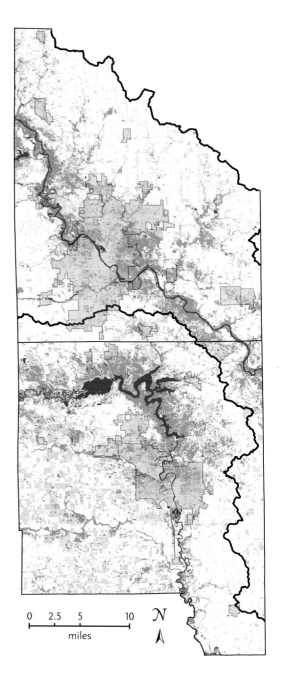

Watershed Boundary
Incorporated Cities
Grassland
Forest
Cropland
Water

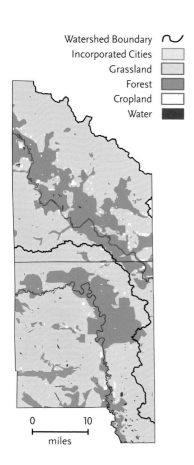

0   10
miles

PLATE 7 (*Above*) Land cover of the
Cedar and Iowa River basins in Linn
and Johnson Counties in the 1840s.
See figure I-2 for location of the
counties. This figure is discussed in
chapter 19. *Illustration by Casey Kohrt,
Iowa Geological and Water Survey,
based on original vegetation interpreted
from General Land Office survey maps,
Anderson 1996.*

0   2.5   5   10   𝒩
miles   ↑

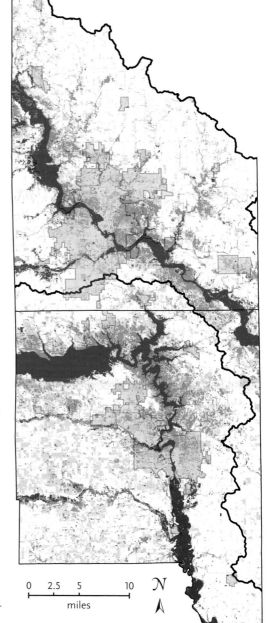

Watershed Boundary ∿
Incorporated Cities ▢
Grassland ▢
Forest ▢
Cropland ▢
Water ◼

PLATE 8 (*Opposite right*) Land cover of the Cedar and Iowa River basins in Linn and Johnson Counties in 2002. See figure I-2 for location of the counties. This figure is discussed in chapter 19. *Illustration by Casey Kohrt, Iowa Geological and Water Survey, based on Landsat satellite imagery from 2002, Kollasch 2008.*

PLATE 9 (*Right*) Extent of flooding in the Cedar and Iowa River basins in Linn and Johnson Counties on June 15, 2008. See figure I-2 for location of the counties. This figure is discussed in chapter 19. *Illustration by Casey Kohrt, Iowa Geological and Water Survey, based on SPOT satellite imagery from June 16–17, 2008, Kollasch 2008.*

0   2.5   5   10   N

miles

∧

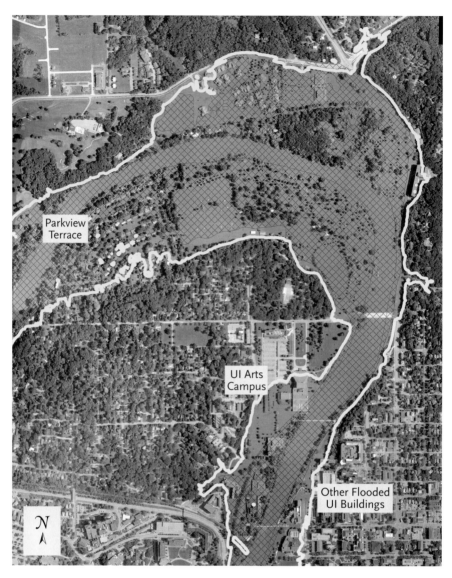

PLATE 10 This map of Iowa City flooding contrasts the 100-year floodplain (red cross-hatch) with the border of the more expansive 2008 flooding (outlined in yellow), which extended well beyond the 100-year floodplain. Note, for example, the flooding outside of expected limits in Parkview Terrace, the University of Iowa Arts Campus, and other flooded university buildings on the river's east bank, all of which experienced considerable and costly damage. This photograph vividly displays that planning for a 100-year flood is planning for an intermediate-sized event, well below what is actually possible. The figure is discussed in the Introduction to section IV. *Illustration by Casey Kohrt, Iowa Geological and Water Survey, based on 2008 flood imagery; flood elevation derived from lidar data (654 feet); and FEMA Digital Flood Insurance Rate Map, 100-year zone.*

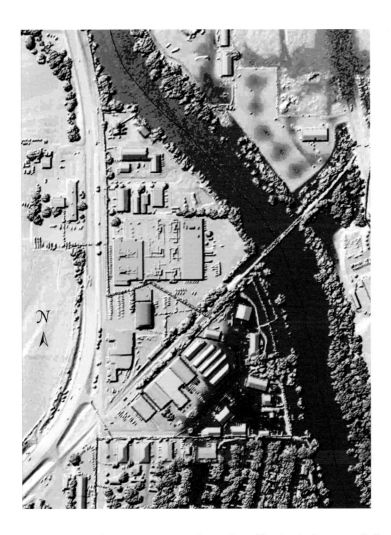

PLATE 11 Research can improve our understanding of floods, which in turn will allow us to better cope with these inevitable natural events. Flood research today is spurred forward by new technological capabilities, such as the use of lidar (which uses pulsing light rays to measure the distance to remote objects). This color-coded image shows the results of a lidar survey of the floodwaters in southern Iowa City, just east of the Iowa City Municipal Airport, performed from an airplane a few days after the 2008 peak flooding. The detailed image of the water's exact distribution will be used to validate numerical computer models that, in future years, will be able to predict the flood extent of a given flood discharge. This information will be immensely valuable for land use planning, insurance maps, emergency flood activities, and the like. Because of the influence of structures, changing flow rates, and many other features on the spread of floodwaters, detailed knowledge about where floodwaters would flow at given river discharges has until now remained elusive. This figure is referenced in the Epilogue. *Image provided by IIHR-Hydroscience & Engineering and the National Center for Airborne Laser Mapping.*

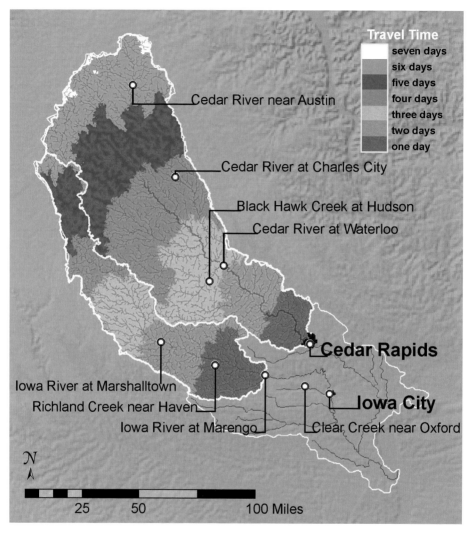

PLATE 12 Map of the average travel time required for streamwater to flow through the drainage network of the Cedar and Iowa Rivers, down to Cedar Rapids and Marengo (just above the Coralville Reservoir on the Iowa River). For the Iowa River, we depict locations only down to the Coralville Reservoir because its human-managed operation influences travel times lower on the Iowa River. See plate 13 for the significance of this travel-time map, plate 1 for locations of Iowa and Cedar River streamgage sites, and figure I-2 for the broader location of the watersheds. This figure is discussed in chapter 2. *Illustration by Radoslaw Goska, based on Mantilla et al. 2006.*

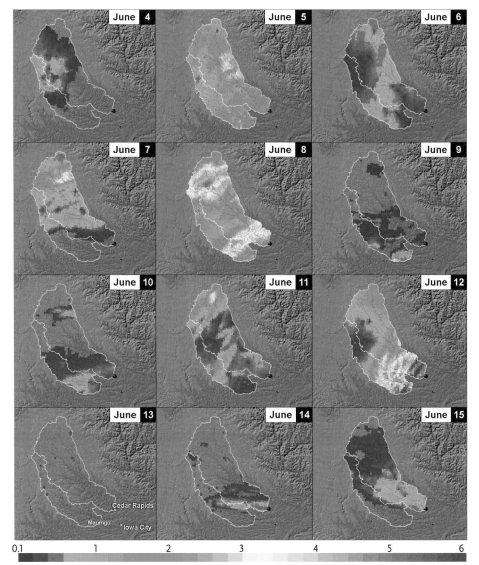

Inches of Rain

PLATE 13 These maps show the amount of rain that fell on the Cedar and Iowa River basins upstream from Cedar Rapids and Marengo throughout the June 2008 flood period. (See figure I-2 to locate these watersheds.) With this knowledge of the location and timing of daily rainfall accumulations and with use of the plate 12 travel-time map, we can explain the 2008 flooding of Cedar Rapids in this way: The lower band of the June 8 intense rainfall reached Cedar Rapids a day or two after that storm, creating the first rise of the big flood. The second, more northerly band of June 8 rainfall required a longer travel time, about five days, and arrived at Cedar Rapids together with the massive rainfall that fell just north of Cedar Rapids on June 12. The compounding of these consecutive storms produced the unexpected, rapid river rise that caused the single, well-defined, and extremely large peak flow in Cedar Rapids and resulted in the worst flooding in Cedar Rapids history. Red dots on the June 13 map are streamgage stations. This figure is discussed in chapter 2. *Illustration by Radoslaw Goska, based on U.S. National Weather Service NEXRAD data, from Hydro-NEXRAD.net.*

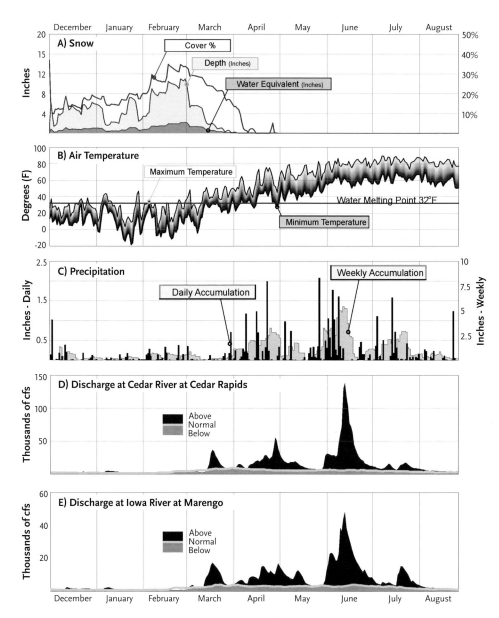

PLATE 14 These five panels summarize much of the story of the 2008 floods. The unusually wet winter deposited an average of 11 inches of snow across the state (panel A, based on daily records from NOHRSC 2008). When air temperatures rose above freezing (panel B, temperatures above Iowa and Cedar River basins, IEM 2008), the snow melted rapidly, causing small floods in March (panels D and E, based on USGS 2009). However, heavy precipitation (panel C, rainfall over Cedar and Iowa River basins, IEM 2008) falling on saturated soils quickly led to additional flooding, first in late April and then spectacularly in June. Note that Panel E shows the river discharge above Marengo rather that at Iowa City, because river flows at the latter are managed by the Coralville Dam. This figure is discussed and references are cited in chapter 2. *Illustration by Radoslaw Goska.*

Michael Burkart

8 The Hydrologic Footprint of Annual Crops

GIVEN THE QUESTION of agriculture's potential contribution to the 2008 floods, one might ask what would have happened had the same weather patterns occurred before the 1830s, when native vegetation still dominated the landscape. Would extreme flooding have occurred then, and if so, would these earlier floods have resembled those of today? Would their extent and magnitude have been as great?

While Iowa's lands certainly flooded in past centuries, floods predating Iowa's agricultural transformation would have differed greatly in character. Iowa today is among the most intensively managed landscapes in the world. This is largely due to its legacy of rich soil, a climate ideal for growing corn, and a culture and economy eager to exploit these resources. These traits have steadily pushed for the replacement of perennial plant cover with annual row crops. The hydrologic footprint of row crop agriculture has been steadily and substantially expanded in parallel with agronomic and engineering changes in the landscape. These changes have systematically eliminated water storage on the land, accelerated the flow of water from the land, expanded the number of streams, and increased the velocity of water moving though streams.

With agriculture's need to remove water as quickly and efficiently as possible, Iowa's rivers now carry more water than before, creating the threat of extreme stream flooding that is magnified in size and extent. No changes have had a more dramatic effect on our land and its hydrology than those associated with agriculture.

Before examining these changes, let's look at Iowa's hydrology before the 1830s when European Americans started transforming the territory to a working landscape. Prior to that time, perennial plant communities, primarily tallgrass prairies, covered most of Iowa and governed the consumption and flow of water through the landscape. The prairie's complex web of dense, diverse grasses and forbs (broad-leaved flowering plants) grew quickly from a dormant state when water and temperatures were suitable, maximizing water consumption throughout the year. Consequently, little water escaped the prairie. Infiltration of water below the prairie root zone was limited, and runoff was generated only during extreme rainfall events or rapid snowmelts. In much of Iowa, runoff from prairies during these periods filled lakes, potholes, and marshes that also supported highly water-consumptive perennial aquatic vegetation. Extreme flood flows were probably rare.

Perennial streams are those that flow year-round and are differentiated from ephemeral streams that flow only during and following rainfall or snowmelt. Runoff is that part of the stream discharge that flows over the land surface and can be observed only during and immediately following rainfall or snowmelt. Perennial streams carry water during long periods between rainfall episodes because groundwater continues to flow into the streams, providing what is called baseflow. In landscapes with a large capacity to store water in surface features, little runoff or baseflow is available for stream discharge.

Since the 1830s, Iowa's native perennial vegetation has systematically been replaced with annual row crops. Prairies, forests, and wetlands have been removed or drained so that by 2002, 62 percent of Iowa's land surface was intensively managed to grow corn and soybeans (U.S. Department of Commerce 2004). While the native perennial vegetation did not allow much water to escape the terrestrial ecosystem, the extensive annual crop system now covering Iowa has been engineered to facilitate movement of water from the landscape. Annual crops inherently contribute to greater stream discharge because they store and consume water during only three or four months of the year; thus much of Iowa's precipitation now falls on land without living plants.[1] A comparison of the average monthly precipitation in Iowa and the growing period of corn and soybeans shows that these crops intercept precipitation only four or five months each year (figure 8-1). In such a landscape, both runoff and baseflow contributions to streams increase (Schilling and Libra 2003), because neither infiltration nor runoff is abated by water-consuming plants between mid-September and mid-May.

A number of factors contribute to today's resulting hydrologic problems. Little overland runoff is impeded by minimal residue from soybeans, and corn residue provides poor water retention when compared to perennial vegetation. Most of Iowa's natural ponds, lakes, wetlands, floodplains, etc., have been drained to increase land for annual crop production. This eliminates the possibility of storing much of today's runoff within the landscape.

The plowing of the prairie also decreased the amount of water-retaining soil organic matter, further reducing the soil's water storage capacity. Early Iowa settlers found a thick column of organically rich topsoil that had formed under

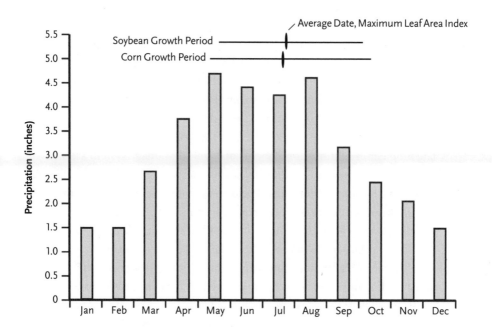

FIGURE 8-1 Average monthly precipitation (inches) at Van Meter in the Raccoon River watershed and annual crop growth periods. Forty-two percent of the annual precipitation falls outside the corn growth period and 50 percent falls outside the soybean growth period. The corn and soybean growing seasons are represented by horizontal lines extending from the average planting date to the average harvest date, with a vertical ellipse near the center marking the average date of maximum leaf area index, a surrogate for the beginning of maximum water demand by the crop. *Illustration by the author, based on data from Hillaker 2009.*

prairies for thousands of years. The prairie's massive root systems accumulated soil organic matter and resisted erosion. In contrast, modern erosion estimates under corn and soybean systems in Iowa using the water erosion prediction program (Foster et al. 1995) model show that as much as 1 percent of the soil's organic matter is lost annually. Today, as a result, less water is stored in the soil, making more available for streams through runoff and baseflow. Few data are available on the specific fraction of organic matter of Iowa's soils prior to the twentieth century, but recent experiments along Bear Creek in central Iowa demonstrate that replacement of annual crops with perennial plant cover quickly restores substantial organic matter. Bharati and others (2002) showed that after only five years of growth, perennial grasses increased soil organic content two- to three-fold compared to soybeans, and four- to seven-fold compared to corn.

Expansion of annual crop production to its current dominance was possible only through direct and extensive manipulation of hydrology via government-supported, engineered conservation practices such as artificial drainage and stream channel modifications. These practices have imposed a large hydrological footprint on Iowa by increasing soil drainage, increasing rates of infiltration, and eliminating most potholes and wetlands. Rapidly removing water from the landscape increased cropland in Iowa but also substantially increased sustained stream discharge and potential flood magnitude and frequency.

Artificial drainage systems were first constructed on a large scale in the late nineteenth century. Components of a typical drainage system are shown in figure 8-2. These systems drain croplands, permitting field tillage earlier in the spring. Subsurface networks of field tiles (gently sloping lines of perforated pipes) buried four to six feet deep collect seeping water and carry it to ditches sufficiently deep to receive water from the tiles. Water also can be fed into the tiles from surface inlets in wetlands, lakes, road ditches, and other areas with standing water. Today, field tiling is becoming more efficient and widespread, and ditches continue to be dredged to accelerate flow and provide lower outlet points for subsurface field drains. Figure 8-3 is a hydrograph of a ditch fed by an artificial drainage system. In this example, the discharge following a one-inch rain is about six times the baseflow. Initially, the increased discharge comes from the depressions that have no surface outlet, closed depressions, through surface inlets. Two days after the initial rainfall, about one-half of the discharge is coming from field tiles. The discharge drops to near its pre-rainfall level after about eight days, when additional small rain events keep the field tile drainage active. Prior to the installation of the drainage system, probably none of the increased discharge would have occurred. All this discharge would have been stored on the landscape, reducing the potential for flooding.

To understand the extent of drainage systems and their tremendous effect on stream discharge, consider the extent of hydric soils, those formed under conditions of saturation, flooding, or pooling long enough to exclude oxygen, in the Raccoon River basin. I have studied these and use them as an example in the following paragraphs. The Raccoon River basin in central Iowa, like most Iowa river basins east of the Missouri divide including those of the Cedar and Iowa Rivers, has its headwaters in the youngest and naturally most poorly drained land surface in Iowa, the Des Moines Lobe (Prior 1991). All Iowa river basins have similar agricultural land-use and climatic trends, supporting the assumption that they all would mimic the Raccoon basin in their discharge and flooding trends.

Iowa's hydric soils have been extensively drained across Iowa to increase annual row crop production. I used soil maps to calculate that in the Raccoon River basin alone, there are more than 728,000 acres of hydric soils, the vast majority of which have been drained. Prior to artificial drainage of these hundreds of thousands of acres, they would have had the potential for storing more than 31 billion cubic feet of water (240 billion gallons) for every foot of water in the closed depressions. This amount of water translates to an average

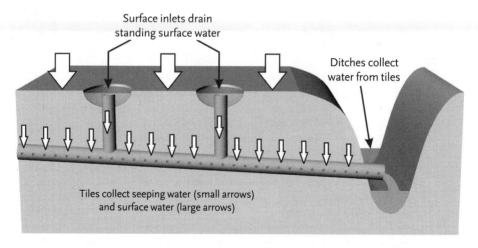

FIGURE 8-2 Diagram of a typical subsurface drainage system including surface inlets that drain wetlands, potholes and lakes; tiles that drain fields; and a ditch sufficiently deep to collect water accumulated in lateral tiles. Lateral tiles are usually 4 to 6 feet under the ground surface. *Illustration by the author and Michael Kundert.*

annual discharge of 1,000 cubic feet per second (cfs). Storage of this amount of water in the watershed's soils would significantly decrease the Raccoon River's average annual discharge. If all this mass of water was stored in springtime, when floods are most common, the reduction in stream discharge during this high-flow season would be even more substantial.

Stream channel modifications of several types have further facilitated water's rapid movement away from Iowa's landscape. In some regions, the number

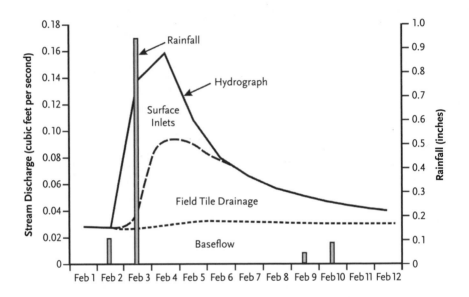

FIGURE 8-3 Hydrograph of a typical tile-drained watershed response to rain when no living plants provide cover for the landscape. The hydrograph plots the quantity of water (vertical axis) against time in days (horizontal axis). The area under the whole hydrograph represents the total volume of water discharged through the stream. The area under the hydrograph and above the dashed line represents water that flows into the stream via surface inlets from the land surface to the subsurface drainage system. The area between the dashed and dotted lines represents the volume of water delivered to the stream by underground tiles that drain the soil (field drainage tiles). The area under the dotted line represents groundwater flow to the stream (baseflow), which shows very little change over the period graphed. A trace of rain (see bars on graph) occurs on February 2, followed by nearly an inch of rain on February 3. The stream discharge increases rapidly on February 2 and 3 as water drains from potholes and other enclosed depressions. On February 3 and 4, water draining from the soil starts to enter the stream. By February 6, three days after the rainfall, the potholes and enclosed depressions are drained, and the stream discharge is supported by water draining from the soil into field tile drains and the fairly consistent groundwater flow. *Illustration by the author, based on Agricultural Research Service, STEWARDS Watershed Database, accessed through James 2008.*

of streams has been increased dramatically: Andersen (2000) showed that the stream network in central Iowa became two to almost six times longer between 1875 and 1972. In the Raccoon River basin, more than 30 percent of the existing stream channels were constructed as a part of artificial drainage systems. New streams and ditches drain large areas directly into rivers; prior to the construction of drainage systems, these areas were drained only through groundwater flow.

Furthermore, destruction of sinuous meanders through stream straightening has shortened natural streams and reduced in-stream water storage. The deepening of channels and ditches creates baseflow by exposing a greater bank area to groundwater discharge. Straightening and deepening combine to accelerate the flow of water, directly increasing flood discharge. These practices also accelerate the erosion of stream banks and streambeds, increasing the deposit of sediment loads in reservoirs. Increased reservoir deposition in turn reduces the flood-storage capacity of reservoirs.

Figure 8-4 documents the result of these many landscape changes. This graphic indicates that increased discharge in the Raccoon River correlates with the increase in the fraction of the watershed covered by annual row crops since

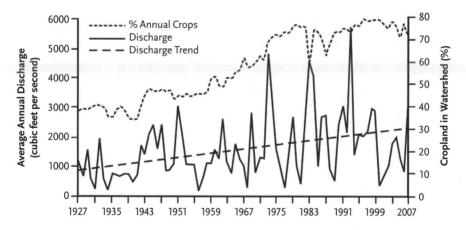

FIGURE 8-4 Annual discharge (flow) in the Raccoon River near Van Meter graphed as the average in cubic feet per second and percent of all land in annual crops in the Raccoon watershed for each year from 1927 to 2007. Note that the average annual discharge follows the increase in coverage by annual crops. The average annual discharge is the total water flowing through the Raccoon River during a year, divided by the number of seconds in a year. The discharge trend resulted from a linear regression. Cropland in the watershed was determined by summing the area in crops for the portions of counties within the Raccoon River basin. *Illustration by the author, based on USGS 2009 and NASS 2009.*

1927. The increase in stream discharge is a result of increases in both runoff and baseflow. The figure shows that discharge has more than doubled during the last 80 years. Note in particular that the peak years have increased at a greater rate than the general trend, with a pronounced increase in peaks beginning in the 1950s through the early 1990s. Some of the increased flow may reflect increases in precipitation, but assuming a linear and conservative correlation, more than 20 percent of the increase in stream discharge can be explained by increases in annual crops. Ongoing research is needed to further define the relative importance of changes in climate, changes in the landscape, and other factors on increased stream discharge.

What steps can be taken to restore the working landscape's hydrological functions and health? How could we increase water storage on the land and slow water's flow to and through waterways? Possible solutions include refocusing crop research, changing land-use policies, and targeting the removal of engineered hydraulic features. Using more perennial crops could expand water storage on the landscape and provide opportunities for exciting and novel research among plant breeders, production-machinery engineers, and commodity production systems engineers. Support for this type of research would have to overcome the political momentum of commodity organizations. Shifting crop subsidies and other economic support from annual to perennial crops would provide a stimulus for innovative farmers.

Iowa's flooding also could be reduced by allowing many streams to once again meander, thus restoring their storage capacity and reducing velocity and sediment load. Repair of damaged and destroyed wetlands could provide multiple benefits to the landscape and watershed, benefits such as water consumption and storage, groundwater recharge, nutrient removal and water treatment, and even potentially harvestable crops.

Technical solutions to stream discharge problems caused by annual crops are relatively well known. Equally well known, however, are the social and political impediments to experimenting with and implementing such solutions. Improved public understanding and pressure will be necessary if farm policies, governmental regulations, and land-use practices are to foster a landscape that generates not only commodity production in Iowa but also flood-resistance and broader environmental health.

Note

1. The average corn planting date in central Iowa is May 9 and harvest date is October 19, with the start of the maximum water demand period (maximum leaf area index) about July 19 (Elmore and Abendroth 2008). The soybean crop is planted, on average, May 24 (Pedersen 2009) and harvested October 8, with a maximum leaf area index date of July 23. The total average time of water use by these crops leaves 42 percent (corn) to 50 percent (soybean) of the annual precipitation to fall on bare soil or soil without living plants to consume or facilitate the storage of water. These percentages are conservative calculations because the annual crop growing season includes a lengthy period of young plant growth between planting and maximum leaf area index (71 days for corn and 60 days for soybean), as well as the period from plant maturity to harvest (during which time the dying plants dehydrate).

References Cited

Andersen, K. L. 2000. "Historical alterations of surface hydrology in Iowa's small agricultural streams." M.S. Thesis, Ames, Iowa: Iowa State University.

Bharati, L., K.-H. Lee, T. M. Isenhart, and R. C. Schultz. 2002. "Soil-water infiltration under crops, pasture, and established riparian buffer in midwestern U.S.A." *Agroforestry Systems* 56: 249–257.

Elmore, R., and L. Abendroth. 2008. "Is all well that ends well? Iowa corn—2008." Iowa State University, University Extension web page. www.extension.iastate.edu/NR/rdonlyres/22E1B296-A473-433A-88F8-7975221B3FA7/92084/1208IowaCorn2008fullreport1.pdf.

Foster, G. R., D. C. Flanagan, M. A. Nearing, et al. 1995. "Hillslope profile and watershed model documentation." In *USDA-Water Erosion Prediction Project: Technical Documentation*, NSERL Rep. 10, ed. D. C. Flanagan and M. A. Nearing, pp. 11.1–11.20. West Lafayette, Ind.: USDA-ARS National Soil Erosion Research Laboratory.

Hillaker, H. 2009. Bureau Chief, Iowa Climatological Bureau, Des Moines. Personal communication, January 2009.

James, D. 2008. Geographic Information Specialist, National Soil Tilth Laboratory, Ames, Iowa. Personal communication, September 2008.

National Agricultural Statistics Service (NASS). 2009. "Iowa crop data by county." Washington, D.C.: U.S. Department of Agriculture. www.nass.usda.gov/QuickStats/Create_County_Indv.jsp. Last accessed January 2009.

Pedersen, P. 2009. "Soybean planting date." Iowa State University, Department of Agronomy and University Extension. http://extension.agron.iastate.edu/soybean/documents/PlantingDate.pdf.

Prior, J. 1991. *Landforms of Iowa*. Iowa City: University of Iowa Press.

Schilling, K. E., and R. D. Libra. 2003. "Increased baseflow in Iowa over the second half of the 20th century." *Journal of the American Water Resources Association* 39(4): 851–860.

U.S. Department of Commerce. 2004. "2002 Census of Agriculture: U.S. summary and county level data," Geographic Area Series 1b [CDROM]. Washington, D.C.: Bureau of the Census.

U.S. Geological Survey (USGS). 2009. "Annual average stream discharge in the Raccoon River at Van Meter." National Water Information System, USGS Surface-Water Annual Statistics for Iowa. http://waterdata.usgs.gov/ia/nwis/annual/?referred_module=sw&site_no=05484500&por_05 484500_1=776911,00060,1,1915,2007&partial_periods=on&year_type=C&format=html_table&date_format=YYYY-MM-DD&rdb_compression=file&submitted_form=parameter_selection_list. Last accessed January 2009.

Wayne Petersen

9 The Hydrology of Urban Landscapes

IN URBAN AREAS, the altered landscape significantly reshapes the hydrologic cycle, which describes the flow of water over and through the earth's surface. Broadly summarized, the cycle follows water from the atmosphere down to the ground as rain or snow. Once on the ground, precipitation may return to the atmosphere through evaporation or transpiration, held by vegetation, or infiltrated into the soil. But it also can rapidly run off the surface of the land-scape, either in a broad "sheet flow" or in a more concentrated, stream-like flow. Floods result when this runoff is excessive and rapid. Whatever route the water follows, it eventually moves on to wetlands, streams, rivers, lakes, and oceans. Ultimately the water evaporates into the atmosphere and starts moving through the cycle once again.

Figure 9-1 shows the fate of precipitation on different types of landscapes. Prior to the 1830s, when native prairies and savannas dominated Iowa, not more than about 10 percent of the annual precipitation became surface runoff. Most of this 10 percent would have been snowmelt or runoff from rain on frozen ground. The other 90 percent was held by vegetation or entered the soil. When rain falls on vegetation, some of it evaporates back into the atmosphere. Some drops to the ground and infiltrates the soil, where it is taken up by plant roots and "breathed" back into the atmosphere (through the process of transpiration). About 40 percent of all rainfall would never have flowed into a stream, because it would have been evaporated or transpired directly back into the atmosphere.

The remaining 50 percent or so of annual rainfall would have infiltrated the soil and slowly percolated downward to feed deep aquifers or become part of the groundwater flow. Because runoff was minimal before the 1830s, surface waters were fed primarily through this slow, steady groundwater flow. Because

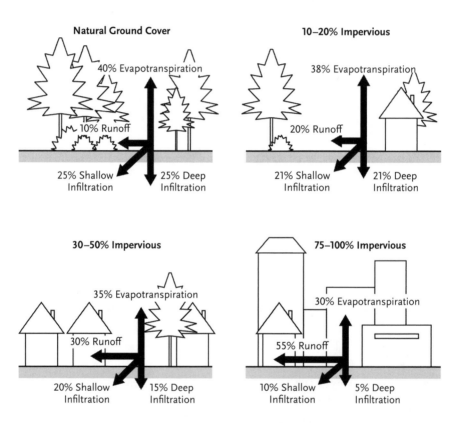

FIGURE 9-1 Note that natural landscapes shed only about 10 percent of precipitation as runoff. As land use intensifies and impervious surfaces cover larger areas, runoff increases and infiltration is significantly reduced. Hydrologic dysfunctionality is manifested most dramatically with the runoff from frequent, small rains. Note that even in significantly urbanized downtown districts, not all rainfall moves as runoff. Because so much annual precipitation comes as small rains (see figure 9-2), a significant amount (up to 30 percent) is held via micro-storage (i.e., puddles) and evaporates. *Illustration from CTRE 2008, which was adapted from Arnold and Gibbons 1996.*

groundwater generally discharges at a constant rate, fluctuations in water levels would have been minimal.

This ancient cycling of water's flow through native prairies and savannas was stable and sustainable. It was infiltration-based and groundwater-driven. The land evaporated, transpired, and infiltrated more rainfall and shed less runoff, certainly compared to today's urban landscapes. Consequently, the land was far more flood-resistant, with water levels in surface waters remaining relatively stable (as well as clean).

Today the stable groundwater-driven hydrology of old largely has been replaced in our urban landscapes by a runoff-driven hydrology, and these landscapes have lost much of their flood resistance. Streams, rivers, wetlands, and lakes now are fed not only by groundwater but also by water running rapidly over the ground surface. When runoff occurs, water levels rise and fall rapidly—a trait referred to as flashiness or flashy hydrology. Flash flooding is a direct result of large volumes of runoff rapidly moving off the landscape. In addition to causing flashy flows and flooding, runoff picks up soil and pollutants as it moves over the ground surface. Runoff from both urban and agricultural lands is a major reason Iowa rivers and streams have poor water quality.

Return to figure 9-1, which diagrams these modern changes in water's flow. The figure shows that as urban development intensifies, runoff increases 20, 30, and up to 55 percent of all precipitation in downtown business districts. Simultaneously, the proportion of precipitation that infiltrates the soil or returns to the atmosphere (through evaporation or transpiration) decreases. This runoff-driven cycling of modern water is characteristic of urban landscapes, where runoff occurs with almost every rainfall event. With these high runoff rates, most urban landscapes are considered "hydrologically dysfunctional"—that is, they don't manage rainfall in a sustainable manner. Hydrologic dysfunctionality is manifested in frequent runoff events that cause spikes in pollutant loads, stream corridor erosion, degraded aquatic habitat, and flash flooding.

Hydrologic dysfunctionality is most obvious with small rainfall events, which account for the majority of Iowa's precipitation and the majority of flashy flows and pollutant loading. Historically, 90 percent of rainfall in Iowa has occurred as events of 1 inch a calendar day or less (figure 9-2) (NWS 2005). Landscapes with good vegetative cover and good soil quality can absorb an inch of rain without generating runoff. But this does not occur much in urban areas. When designed with hydrological health in mind, modern landscapes—even urban landscapes —should be able to manage an inch of rain without generating runoff. The types of design features that would support this goal are discussed in chapter 23.

Why do urban landscapes generate so much runoff? A key feature is their many impervious surfaces—parking lots, streets, roofs, driveways, and the like. Impervious surfaces prevent rain from infiltrating the soil, and instead generate runoff. Refer once again to figure 9-1, which shows how impervious surfaces increase as urban development intensifies. (Note, however, that even in the most highly urbanized settings not all rainfall moves as runoff. Impervious surfaces such as parking lots catch rain in puddles, hold it, and allow it to evaporate when small rains occur.)

Impervious surfaces are usually engineered to drain runoff quickly, moving it rapidly to nearby streams. Rain falling on roofs and driveways is typically routed

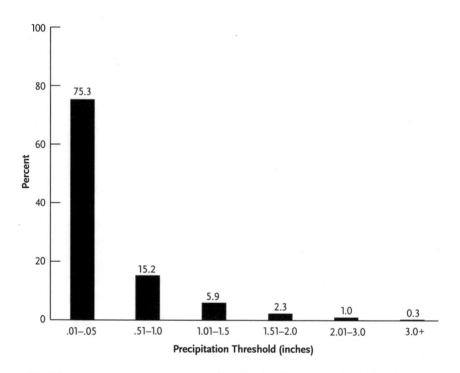

FIGURE 9-2 Frequency of 24-hour precipitation events, Davenport, Iowa, 1948–2004. Note that historically, 90 percent of rainfall events are less than 1 inch per 24 hours. Natural landscapes would not generate runoff with 1 inch or less of rain; that amount of rain would be evaporated, transpired, or infiltrated into the soil. Runoff is generated, however, whenever rain falls on impervious urban surfaces, which causes flashy streamflows, water pollution, and increased flood potentials. Rainfall patterns throughout Iowa are similar to these charted for Davenport. *Illustration by Jason Johnson, U.S. Natural Resources Conservation Service, from rainfall data compiled by Ray Wolf, 2005.*

directly to the street, concentrated in curbs and gutters, and then collected in storm sewers that quickly discharge storm runoff into a stream. Direct connections between impervious surfaces facilitate the shedding of runoff. Disconnecting a downspout from an impervious driveway (i.e., diverting downspout flow into a rain garden or a yard with good soil quality) helps reduce the volume of runoff that reaches storm sewers and receiving streams.

Even green spaces (i.e., lawns) in urban areas often act as impervious surfaces and generate runoff, due to soil quality degradation. This is a major contributor to hydrologic dysfunctionality. Soil quality degradation typically results from land-disturbing activities (grading) associated with development. Topsoil is often stripped and hauled off site in the first phase of development. Mass reshaping follows, to make the landscape conform to the design of the new development. This grading activity alters the natural soil profile and leaves compacted subsoil behind. In fact, landscapes are often intentionally compacted to facilitate the shedding of water (as runoff) away from buildings. Sometimes a layer of topsoil is spread over the compacted subsoil after grading is completed, but this is not always done. Either way, laying sod over these altered and compacted urban landscapes yields green space that can't infiltrate rainfall as well as the natural, healthy soil once did.

The typical result is green space that sheds runoff after a very small amount of rain is absorbed. If you live in a new subdivision, walk around your yard during a period of extended or heavy rain. Don't be surprised if your footsteps cause splashing as you walk across the grass. Look closely and you'll see that water is running off your lawn rather than being absorbed into the soil. You often see lawns "oozing" water into the gutter of streets after a rainfall event. Rain has stopped falling, the sun is out, and streets have dried up—except for wet gutters where saturated lawns are oozing water that could not infiltrate (figure 9-3). You sometimes see this same phenomenon on lawns that are being irrigated. A lawn oozing water into the street gutter is a tell-tale sign of a hydrologically dysfunctional landscape.

I believe that urban landscapes contribute more to peak runoff events than we currently think, but no one really knows how much. What percent of the 2008 flood peak that slammed into Waterloo or Cedar Rapids came from urban or road runoff? I am not aware of data that answer this question.

However, it is hard to conclude that urban runoff played a major role in the 2008 record flooding of eastern Iowa's major river basins. These watersheds are dominated by row crop agriculture. Runoff from agricultural land had to be the primary source of the runoff that caused the flooding. A quick analysis

FIGURE 9-3 Water oozes from a hydrologically dysfunctional landscape featuring sod laid over compacted subsoil. This photo was taken a couple of hours after 2 inches of rain fell over about a two-hour period of time. After the storm passed, the sun came out and the street dried up—except for the wet gutter where the water continued to ooze out of the sod. *Photograph by the author.*

of the 3.6 million acres in the Cedar River basin above Cedar Rapids indicates that in 2002, only about 4 percent of the basin was in road, residential, or commercial/industrial land use. Row crop agriculture (corn and soybean production) accounted for about 73 percent of the basin. About 18 percent was in perennial grass or forage cover, and the remaining 5 percent was in woodlands (IDNR 2008).

But in smaller watersheds, like the Fourmile Creek watershed in Polk County, urban runoff plays a definite role in localized flooding. This watershed, with its 76,624 acres, is a small fraction of the size of the Cedar River watershed. However, about 36 percent of the Fourmile Creek watershed is urbanized (Welch 2008), much more land proportionately than the urbanized land in the Cedar River basin above Cedar Rapids.

Fourmile Creek originates in Story and Boone Counties. It drains the agriculturally dominated north-central part of Polk County before it takes on runoff from eastern Ankeny and moves into the Des Moines metropolitan area. These two areas have been undergoing rapid urbanization, and the area will continue to grow. New development will increase even more the percent of imperviousness in the Fourmile Creek watershed. More imperviousness will generate larger volumes of runoff, and the potential for larger and more frequent floods will increase. This general problem is a concern in rapidly developing areas throughout Iowa, including Iowa City and Cedar Rapids.

Fourmile Creek has a history of frequent flooding in its lower reaches (the Des Moines area). Residents of the lower watershed are concerned about ongoing urbanization in the upper watershed. These concerns are valid. Problems with flooding will be aggravated downstream in the watershed, unless new green strategies that lead to more sustainable urban growth are implemented as the watershed develops.

There's reason to be optimistic though. Ankeny has commissioned a study to investigate ways to incorporate "green" features into new growth and retrofit built-out areas with stormwater wetlands, bioretention, and stream corridor stabilization and buffering. A watershed protection project that addresses both agriculture and urban runoff is also under way for Fourmile Creek. These efforts should help reduce flood potentials in the Fourmile Creek watershed, while enhancing water quality as well as the ecology and recreational value of the Fourmile Creek corridor. Green growth and ecological enhancement are compatible and can improve the quality of life in Iowa's growing communities.

While urban runoff may not have been the major contributor to the 2008 flood peaks in eastern Iowa, we know that urban runoff contributed to the peaks. And we must anticipate more problems with localized flash flooding along streams like Fourmile Creek in the future, unless we embrace new strategies for more sustainable growth. Urban growth, combined with the potential for more frequent, intense storm events as weather patterns change, makes managing urban runoff in more sustainable ways a major goal for the future.

References Cited

Arnold, C. L., and C. J. Gibbons. 1996. "Impervious surface coverage: The emergence of a key environmental indicator." *Journal of the American Planning Association* 62(2).

Center for Transportation Research and Education (CTRE). 2008. *Iowa Stormwater Management Manual*, Section 2B-1. Ames, Iowa: Iowa State University, CTRE. www.ctre.iastate.edu/PUBS/stormwater/index.cfm.

Iowa Department of Natural Resources (IDNR). 2008. IDNR's Watershed Initiative,

interactive mapping website. www.igsb.uiowa.edu/nrgislibx/watershed/
watersheds.htm. Accessed December 2008.

Welch, Jennifer. 2008. Urban Conservationist, Polk County Soil and Water
Conservation District. Personal communication, December 2008.

Wolf, R. 2005. "Daily surface weather observations for Sioux City, Iowa and Moline,
Ill. 1948–2004." Davenport, Iowa: National Climatic Data Center, NOAA.
http://lwf.ncdc.noaa.gov/oa/ncdc.html.

John Castle

10 The Coralville Dam and Reservoir
Design and Operation

~~~~~~

THE CORALVILLE DAM was constructed in the 1950s by the U.S. Army Corps of Engineers (Corps) and put into operation in September 1958. The earth-filled dam is 1,400 feet long, 100 feet tall, and 650 feet wide at the base.[1] The dam created the Coralville Reservoir, a large body of water that under normal conditions—when the water surface is at 683 feet elevation above mean sea level (msl)[2]—reaches 21.7 miles upstream, has a surface area of about 5,400 acres, and holds 28,400 acre-feet of water.[3] Inflowing sediments, primarily from agricultural land, are gradually reducing the reservoir's depth and water-holding capacity: by 2008 the reservoir's total storage capacity was 14 percent less than when newly constructed (Martin 2008).

At the dam's east end stands the water control structure: three vertical slide gates, each 8.33 feet wide by 20 feet tall, that control the amount of water released from the reservoir back into the Iowa River. A 500-foot-wide emergency spillway lies at the dam's west end. The emergency spillway crest or top is 712 feet, 31 feet lower than the top of the dam. The dam is designed to safely handle flows over the emergency spillway as large as 244,000 cubic feet per second (cfs). Downstream flooding can result either from large but regulated water release through the dam's gates or from uncontrolled spillway overflows. Upstream lies the Hawkeye Wildlife Area, land owned by the Corps and leased to the Iowa Department of Natural Resources that is part of the flood storage area. Lake Macbride, an impounded lake within Lake Macbride State Park on the east side of the reservoir, is also considered part of the reservoir storage area when water in the lake is above the 712-foot level, the height of the spillway separating Lake Macbride and the reservoir.

The primary purpose of the dam and reservoir is reducing flooding on the Iowa River and, by storing Iowa River floodwaters, ultimately reducing flooding on the Mississippi River (which is fed by the Iowa River 83.3 miles downstream from the dam). As was demonstrated in 1993 and again in 2008, the dam can reduce flooding downstream but cannot eliminate all flooding, nor was it ever intended to do so. The dam's secondary purposes are low flow augmentation (water storage and release during dry conditions), water quality improvement (by allowing sediment to settle out), provision of fish and wildlife habitat, and recreation.

The normal reservoir level varies with the time of the year. In summer the reservoir is maintained at elevation 683 feet. The reservoir is raised to elevation 686 feet in October to provide additional water surface area for migrating waterfowl. In December, the level is returned to elevation 683 feet. In February, the reservoir level is lowered to elevation 679 feet in anticipation of spring rainfall. This adds a small measure (3 percent) of additional flood storage capacity. The reservoir is returned to elevation 683 feet by Memorial Day. Once the desired level for any given period is reached, water's inflow (from upstream tributaries) and outflow (through dam gates) is balanced to keep the reservoir level constant.

The dam is managed by the Water Control Section of the Rock Island District office of the Corps, with input from personnel at the Coralville Reservoir office (who carry out the desired operations). Management decisions are guided by a detailed regulation schedule (Corps 2001). The schedule specifies limits to release rates from the Coralville Dam according to the time of year, elevation of the reservoir, downstream level of the Iowa River, and even the level of the Mississippi River at Burlington. For example, from May 1 to mid-December, the reservoir is to be held at elevation 683–686 feet by using release rates between 1,000 and 6,000 cfs, unless well-defined flood conditions exist downstream. (In that case, a variety of other specific discharge rates are mandated.) Alternatives are also given for major flood and major drought conditions.

When flood conditions exist, the dam's release rates are lowered to reduce

Iowa River flood crest levels expected at downstream streamgages near Lone Tree and Wapello. Flows on the Cedar River, which joins the Iowa River at Columbus Junction, also help determine dam releases. Often the Cedar River flows are high and the release rates at the dam are reduced to try and keep downstream streamgage readings within the schedule's limits.

The schedule is not an absolute guide, but it is followed unless circumstances dictate otherwise. Deviations only are considered for compelling reasons as they must be authorized by the Mississippi Valley Division office of the Corps, the next level of command above the Rock Island District. At some point in the spring of 2008, both schedule restrictions and such deviations were used to determine the amount of water that was released from the reservoir.

The basic regulation schedule has been in effect since 1964 and came about after minor flooding of farmland below Iowa City in 1959. Negotiations and a lawsuit brought by downstream agricultural interests resulted in lowered limits for water release during the growing season. The schedule is still focused primarily on reducing the flooding of agricultural land. Major changes to the schedule were made in 1991–1992 (raising the summer level to elevation 683 feet and raising the release limits at elevations above 707 feet, in order to compensate for storage loss due to sediment accumulation in the reservoir).

The management of potential flood flows from the reservoir always involves trade-offs that balance the needs and protection of one location with those of another location. For example, releasing large amounts of water early during flood years may reduce the peak of very large flood flows later on, but the early releases are certain to flood the lowest-lying areas downstream—perhaps unnecessarily. Because reductions in release rates at the dam are focused on controlling river levels downstream from Iowa City, release rates frequently are lowered even though there is no high water in the Iowa City area.

Since no one can predict the weather, decisions about such trade-offs are difficult to make. The regulation schedule was developed using models and historic data to arrive at the best balance for most conditions. With all the changes in the downstream floodplain since the 1960s, the current plan may or may not remain the best, but major changes to the plan would require a comprehensive study partially paid for by downstream interests. To date, no agencies or interests have been willing to share costs with the Corps on such a study.

How did these many stipulations and application of the regulation schedule play out during the unusual hydrological events of 2008? With the previous winter's heavy snow and the following spring rains, the reservoir level was near the summer elevation of 683 feet already in early March.[4] For three days during the third week of March, the dam's release rate was lowered to near the minimum of 1,000 cfs because the Iowa River level at Wapello had exceeded the limit of 22 feet. The high river levels at Wapello resulted from high Cedar River inflows.

During this time, inflows into the reservoir were high and the reservoir level increased to elevation 703 feet. Thus, after three days of minimum discharges, the release rate was raised to 10,000 cfs, and the reservoir level slowly began to drop.

The reservoir had dropped to 694 feet when, in the third week of April, release rates from the dam were again lowered to near 1,000 cfs because the Mississippi River stage at Burlington was predicted to rise above 18 feet. This time the release rate was lowered for 7 days, and the reservoir level again rose to elevation 706 feet. Even though outflows were returned to 10,000 cfs at the end of 7 days, high inflows caused the reservoir level to continue to rise to 709 feet. Once the reservoir elevation exceeded 707 feet, releases from the dam were guided by the major flood section of the regulation schedule.

After May 1, releases from the dam would normally be limited to 6,000 cfs. However, the reservoir elevation was still above 707 feet, so the reservoir continued to be operated under the major flood schedule. Releases were maintained at 10,000 cfs until May 21, when the reservoir elevation fell below 707 feet. Releases were then reduced gradually to 6,000 cfs. After discussions between the Coralville Reservoir office and Rock Island's Water Control Section, it was decided that since wet soil conditions from rainfall would prevent field work along the river anyway, the dam's outflows should be raised to near 10,000 cfs in an effort to increase the reservoir's storage capacity. Authorization for the deviation from the schedule was received from the Corps' Mississippi Valley Division office.

On June 5, as heavy rains fell upstream and the projected reservoir levels continued to rise, releases from the reservoir were increased above 10,000 cfs. Downstream impacts begin at releases of 10,000 cfs. The River Front Estates housing development below the dam was affected first, with the Edgewater area in Coralville following at about 12,000 cfs. The northbound lanes of Dubuque Street were covered at about 13,500 cfs, as were individual locations in the Parkview Terrace area. Several times increases in release rates were delayed at the request of Iowa City, the University of Iowa (UI), and Coralville to allow more time to carry out flood-fighting efforts. On June 10, the control gates were completely opened, and the release rate was increased to nearly 20,000 cfs. Later that evening, the reservoir waters began to overtop the emergency spillway. Reservoir water continued to flow over the spillway until June 24.

Runoff from heavy rains upstream continued to push the reservoir level higher and increase the flow rate over the spillway until the reservoir crested at 717.02 feet on June 15 (figure 10-1). At that point, the water flowing over both the dam's emergency spillway and Lake Macbride spillway was five feet deep. While it is true that once water overtops the spillway control of the flow over the spillway is lost, still adjustments can be made to the controlled water release through the dam gates. Total downstream releases can be reduced for short periods of time to reduce the effects of flash flooding or for other emergencies, but these reductions will later increase the uncontrolled flows over the spillway.

Maximum inflows to the reservoir from the Iowa River in 2008 were estimated at 57,000 cfs (Stiman 2008). This compared to a maximum inflow of 41,000 cfs (Koellner 1993) in 1993. Despite the difference in inflows, the maximum reservoir elevation was almost identical in 1993 (716.72 feet) and 2008

FIGURE 10-1 Coralville Dam and emergency spillway (on the left) at flood crest, June 15, 2008. The outflow gates are on the dam's right. *Photograph by Chad D. Andrews.*

(717.02 feet). The 717 foot-level is significant because the Corps owns land or has bought easements to permit flooding up to 717 feet. Any flood damage caused by reservoir levels above 717 feet would be the responsibility of the Corps. In 2008 (and in 1993), conditions were one good rain away from causing serious damage upstream in the Amana Colonies and at the Amana Refrigeration Plant and additional damage below the dam. At its 2008 maximum, the water was five feet deep going over the emergency spillway, and the reservoir surface had increased from the normal 5,400 to 28,200 acres and from 21.7 to nearly 45 miles long (Martin 2008).

How did water releases during the 1993 floods compare to releases in 2008? In 1993, water releases were reduced in order to protect the Iowa City and UI water treatment plants. Total releases from the dam gates and spillway overflow were restricted to 24,600 cfs. This was accomplished by leaving the control gates partially closed. Thus downstream flooding was less in 1993 than it would be in 2008, but this lessening was carried out at a risk: Had additional significant upstream rains fallen during the period of lower outflow, the reservoir may not have had the reserve capacity to handle the surge of inflowing water.

Fortunately the weather cooperated, and 1993 inflows remained lower than those of 2008. Also, by 2008, Iowa City had a new water treatment plant and the UI water treatment plant had greater protection. Thus, in 2008, the dam gates were left completely open (figure 10-2), and downstream flooding was magnified—a reality forced by the tremendous upstream rains and resulting very large inflows into the reservoir. The outflow maximum (including both dam gate and spillway flows) was 39,900 cfs (see table 7-1) on June 15. Flows at the Iowa City streamgage were even higher because of additional input from Rapid Creek and Clear Creek, which enter the Iowa River between the dam and Iowa City. As large as these figures are, the maximum inflow was far greater than the maximum outflow, demonstrating that the reservoir did indeed lower the peak of downstream flooding. With the dam gates fully open and total outflows so high, the reservoir flowed over the spillway for only 14 days in 2008. (In 1993, water flowed over the spillway for 28 days, beginning on July 5.) The dam gates were left completely open in 2008 until July 6, when the reservoir elevation was again below 700 feet.

The 1993 and 2008 floods also differed in how they were handled. Technological advances since 1993 have allowed the National Weather Service to better predict rainfall. In 1993, predictions of flow rates on the various rivers could not be made until rain had actually fallen, so decisions on reservoir release rates had to be delayed. In 2008, the accuracy of predictions had increased to the point that inflow estimates included rainfall predictions 24 hours into the future. This allowed for releases to be increased earlier and helped conserve storage capacity of the reservoir.

Communication among various agencies involved in the flood fight was much more open in 2008. Teleconferences involving the Corps and representatives of Iowa City, Coralville, the UI, Johnson County, and the Iowa Department of Transportation were held every morning and often again in the afternoon. Weather conditions and forecasted changes in release rates were presented by the Corps. The probable effects of those changes were discussed.

FIGURE 10-2 Even when the dam's emergency spillway is overtopped, downstream flood-waters can be controlled to a degree by varying the release of water through the dam gates. This dramatic photograph shows water flowing from the reservoir into the dam-gate intake on June 15, the day the reservoir crested. The gates were opened wide; maximum outflow through the gates reached 21,000 cfs. While these large dam-gate releases combined with emergency spillway overflows to raise the level of floodwaters in Iowa City, they prevented the reservoir from rising even higher and endangering the upstream Amana Colonies. The larger dam-gate releases also drew down the reservoir faster, reducing the possibility of later, additional flooding. In 1993, dam-gate releases were kept below this level in order to protect the Iowa City water treatment plant. *Photograph by Kaylene F. Carney, U.S. Geological Survey.*

One constant throughout the years has been the safety of the Coralville Dam itself. At no time in either 1993 or 2008 was there any concern about its stability. The Coralville Dam is built on a very stable foundation, and no movement has been detected over the years. Thorough inspections of the dam after both major floods verified that the floods had no affect on the safety of the dam.

Notes

1. Information found throughout this chapter on the structure and operation of the Coralville Dam and Reservoir has been taken primarily from Corps (1991).

2. Mean sea level (msl): the elevation of the reservoir surface above average sea level. All elevations in this chapter are elevations above msl.

3. The usual way to describe the reservoir's capacity, an acre-foot of water, equals the amount of water required to cover an acre of land to one foot depth. One acre-foot equals 1,234 cubic meters of water.

4. Information on the operation of the dam and associated flow levels during the 1993 and 2008 flood periods, as reported in this chapter, is recorded in Corps 1993 and 2008, unless otherwise indicated.

References Cited

Koellner, William. 1993. Chief, Water Control, U.S. Army Corps of Engineers, Rock Island District Headquarters. Personal communication, 1993.

Martin, Dave. 2008. Water Control Engineer, U.S. Army Corps of Engineers, Rock Island District Headquarters. Personal communications, August–November 2008.

Stiman, Jim. 2008. Chief, Water Control, U.S. Army Corps of Engineers, Rock Island District Headquarters. Personal communication, July 16, 2008.

U.S. Army Corps of Engineers (Corps). 1991. *Water Control Plan with Final Supplement Environmental Impact Statement, Coralville Reservoir, Iowa.* Rock Island, Ill.: U.S. Army Corps of Engineers, Rock Island District.

———. 1993, 2008. *Daily Hydraulics Log Book, Coralville Lake.* Johnson County, Iowa: U.S. Army Corps of Engineers, Rock Island District, Coralville Lake office.

———. 2001. *Water Control Manual, Coralville Lake, Table C-2.* Rock Island, Ill.: U.S. Army Corps of Engineers, Rock Island District.

Robert F. Sayre

11 The Dam and the Flood
Cause or Cure?

"FORTUNATELY," IOWA CITY historian Irving Weber wrote in 1985, "since the Coralville Dam and Reservoir have been installed, the Iowa River has been under control and the area no longer suffers disastrous floods."

The floods of 1993 and 2008 have since proven Weber wrong. He also might have known better. The Coralville Dam, with a storage capacity of 435,300 acre-feet (when the reservoir is at 712 feet elevation at the crest of the emergency spillway), was designed to hold only 2.6 inches of runoff from the Iowa River's entire 3,115-square-mile drainage area above the dam (Corps 1991-a). When full, and when the dam gates are wide open, water is released at upwards of 20,000 cubic feet per second (cfs), more than enough to flood low-lying sections of Coralville, Iowa City, and the University of Iowa (UI) campus.

Four floods that Weber listed at the end of his article were this big or bigger.

A flood in 1851 reached an estimated flow rate of 70,000 cfs, and a flood in 1881 had reached 51,000 cfs. Weber also cited precise records (kept since 1903) of the June 8, 1918, flood that reached 42,500 cfs and the June 17, 1947, flood that reached 33,800 cfs.

By comparison, the maximum flow rate at Iowa City of the 1993 flood, reached on August 10, was 28,200 cfs, and the 2008 flood, on June 15, was 41,100 cfs (USGS 2008).

Weber was not the only person to assume that the Coralville Dam had eliminated the danger of disastrous floods on the Iowa River. In 1959, a year after the Coralville Dam went into operation, the Iowa City City Council approved the development of Parkview Terrace subdivision, accepting developers' arguments that the dam would now protect this low land adjacent to the city's lower City Park (Linder 2008). In the 1960s the university proceeded with enlarging the Arts Campus and building the English-Philosophy Building, also in the floodplain. Since then even more UI buildings have gone up there, as have commercial and residential buildings in Coralville and Iowa City. Flood control dams, it has been shown, may actually increase flood damage. First, they encourage development in floodplains. Then, when there is more water than the reservoir can hold, damage is greater because there are more structures to be damaged than there were before. At the worst, as happened in Rapid City, South Dakota, in June 1972, a dam fails and many lives are lost (Rahn 1975).

All actors in Iowa's 1993 and 2008 tragic floods, we can see now, were living with a false sense of security. They did not realize (or preferred to ignore) important facts about the history and purposes of the Coralville Dam, its limitations, and the policies of the U.S. Army Corps of Engineers (Corps) in managing it.

The Flood Control Act of June 28, 1938, which authorized the Coralville Dam, was passed in the wake of the disastrous Ohio River flood of 1937. The Corps recommended building eleven reservoirs on the tributaries of the Mississippi and Missouri Rivers to prevent their waters from adding to another disaster on the lower Mississippi (Corps 1991-a, "Pertinent Data"). Thus the Coralville Dam's initial purpose was reduction of flooding on the Mississippi, although that has come to include reducing floods on the Iowa River. Still, its operations have never been directed locally but instead are controlled by the Corps District Headquarters in Rock Island, Illinois. There the Corps also directs the Saylorville and Red Rock Dams on the Des Moines River, the navigation locks and dams on the Illinois River, Lock and Dams 11–22 on the Mississippi, two dry reservoirs in Illinois and Wisconsin, and many levees (Martin 2008).

In addition to flood control, the four priorities guiding its operation are, in order, low flow augmentation (providing sufficient water for sanitation and wildlife during droughts); water quality (principally by allowing sediments and pollutants to settle above the dam); fish and wildlife; and recreation.

An additional consideration in the operation of the Coralville Dam is the condition of the Cedar River. Although nominally a tributary of the Iowa, the Cedar is much larger, draining 7,870 square miles—more than twice the drainage area of the Iowa above the dam—but it has no flood control dam. Thus when there is danger of flooding below their meeting point at Columbus Junction, Corps policy is to hold back water on the Iowa until the crest of the Cedar has had time to pass through.

Guiding the Corps in operating the dam are a number of lengthy government documents, principally the *Emergency Plan for Coralville Dam and Lake* (Corps 1984); *Regulation Manual for Coralville Lake* (Corps 1991-a); and the *Water Control Manual, Coralville Lake* (Corps 2001). They are large loose-leaf notebooks, some with a hundred pages or more of text, facts, and tables and many more pages of maps, engineering drawings, graphs, and other supporting documents. They are daunting and, to a layman like me, almost overwhelming.

Several general comments can be made about them, however. First, one wonders whether they are up to date. Both the *Emergency Plan* and *Regulation Manual* predate the 1993 flood, which one would think should have necessitated changes. Yet the very size and detail and complexity of the documents speak to the difficulty and expense of compiling them. The *Environmental Impact Statement*, a supplement to the *Water Control Plan* (Corps 1991-b), required over a year of study, data collection, and public hearings to which all individuals, landowners, public bodies, and businesses that might be affected were invited. Policy for operating the dam and reservoir is difficult to arrive at. It affects lots of people. It must cover all the purposes. And it must consider all the ranges of midwestern weather.

The occasion for the *Environmental Impact Statement* was a plan by the Corps to raise the summer and winter conservation pool elevation from 680 feet to 683 feet above mean sea level.[1] This conservation pool's elevation is the reservoir's target level, which varies with the season in order to provide for recreational use, sanitation, and wildlife in periods of drought (see chapter 10). The increase was sought so as to compensate for the sediment that had collected since 1958. Since this rise of three feet would reduce the reservoir's reserve capacity and thus its flood control capability, the plan also called for starting emergency release (technically called intermediate magnitude flood regulation) at elevation 707 instead of 710 feet (Corps 2001).

To put these numbers in context, when the water level reaches elevation 712 feet it overtops the emergency spillway and can result in a serious flood, like those of 1993 and 2008. One also has to know that the capacities of the reservoir at elevations 707 feet and 712 feet are 300,000 acre-feet and 421,000 acre-feet, respectively. Thus the reserve capacity at 707 feet is only 121,000 acre-feet, a reserve that can be rapidly filled if water is flowing into the reservoir much faster than it is flowing out, as is common in the time of a flood.

A solution would be to begin emergency release earlier, say at elevation 690 or 700 feet. A graph prepared by the Corps (figure 11-1) shows that at elevation 690 feet the reservoir holds approximately 70,000 acre-feet, which leaves a reserve capacity of 351,000 acre-feet. At elevation 700 feet it holds 180,000 acre-feet, which provides 241,000 acre-feet of reserve capacity. While these reserves are not sufficient to protect against a really large and rapid inflow, they would provide a greater margin of protection. Computer models done after the 1993 flood show that under some conditions, having the reservoir at a lower level and releasing water sooner might have prevented the water from going over the spillway (Martin 2008).

There are two problems with this solution, both resulting from our inability to predict the weather. First, if the weeks and months following a flood scare become very dry, there will be less water available for low flow augmentation, recreation, and fish and wildlife. But since the minimum outflow for them is only 150 cfs, this possibility seems remote. The more serious problem is that releasing extra surges of water earlier will flood some low-lying riverside areas that otherwise would have remained dry, places such as Edgewater Drive in Coralville, River Front Estates in River Heights north of Iowa City, and farmland between Iowa City and Columbus Junction. Channel capacity below the dam is stated to be 8,500 cfs, and so releases of 10,000 cfs or more damage these homes and flood riverside cropland (Corps 2001).

The water control documents recount many controversies over this issue, illustrating that management of a reservoir requires deciding which landowners' needs will be met and which landowners will be flooded to protect others. In 1959 and again in 1960, less than two years after the dam went into operation, "landowners objected to the high growing-season release rate of 8,500 cfs" because it flooded their downstream property. By 1968 there was much more development along the shore of the reservoir itself, and "recreational users wanted higher pool levels . . . for boating and more desirable aesthetic conditions." In 1982, the "spring conservation pool was changed from elevation 670 to 675 and the drawdown date was delayed from 1 February until 15 February"—both of which resulted in higher reservoir levels and lower release rates. Modifications in later years went further (see current levels in chapter 10). In 1999, when the Corps recommended an increase in the maximum summer release rate from 6,000 to 8,000 cfs, the Iowa Department of Natural Resources and downstream landowners and communities all objected, and the release rate was returned to 6,000 cfs (Corps 2001).

Surprisingly, nowhere in the 2001 *Water Control Manual* is there mention of requests from the University of Iowa or residents and business owners in Iowa City or in Coralville for higher, earlier release rates that might have prevented or lessened the damage from greater floods like those of 1993 and 2008. Until 1993, residents and administrators appear to have been confident that no such floods would occur. After 1993 they put their faith in measures such as higher levees, flood gates, and automatic pumps. It seems reasonable to suggest that the university, Iowa City, and Coralville should now request that the Corps revise its water control plans and that funds be appropriated for new studies and hearings.

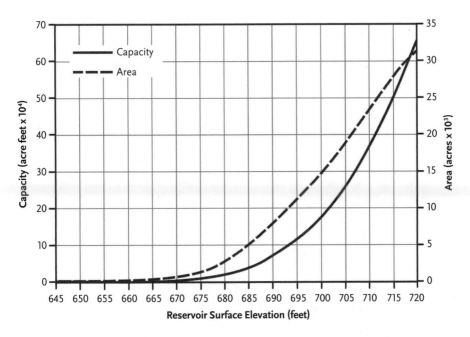

FIGURE 11-1 Capacity and area of the Coralville Reservoir at different elevations of the reservoir's surface, from a 1999–2000 survey. The volume of water stored (left scale and solid line) rises rapidly as the area covered by water (right scale and interrupted line) increases. For any elevation of the reservoir, this graph allows one to determine its volume and surface area. Reserve capacity can then be calculated by subtracting the volume at a certain reservoir height from 412,000 acre-feet, the volume at elevation 712 feet, at which height the emergency spillway is overtopped and serious downstream flooding begins. Earlier release of larger quantities of water keeps the reservoir's volume and surface area lower, thus preserving a larger reserve capacity. *Illustration adapted from Corps 2001, plate 2-2.*

To argue in favor of any flooding is unpopular. To favor flooding one area in order to protect another can be very unpopular, the equivalent of asking protection for yourself at the expense of a neighbor. But sooner or later society will have to decide what degrees and locations of flooding are more acceptable than others. Is it wise to protect cropland (some of it very marginal) and a few houses and buildings from more numerous small floods at the risk of having much larger floods that endanger hundreds of houses and buildings that cost much more and would be far more expensive to repair?

One way out of these difficult choices would be to acquire the low-lying, most-at-risk properties (figure 11-2), as FEMA and other agencies already do in other flood-prone areas. Some property owners would be bought out. Others, like owners of agricultural land, might sell flood easements. In these ways the owners of most-at-risk property would be compensated, while owners of

FIGURE 11-2 Flood-damaged house on low-lying land, close to the Iowa River on Edgewater Drive, Coralville, November 1, 2008, four and a half months after the June flood. Such properties are in frequent danger. Their purchase and demolition could permit safe, earlier releases of larger volumes of water, thus protecting other less-at-risk properties from possible damage later by larger floods. *Photograph by the author.*

property that is less at risk would receive greater protection. The buy-outs and easements would not eliminate all possible flood damage. But they would reduce its expense, and when accompanied by other measures such as better land management practices in the upstream watershed, resulting in better water conservation, they would permit earlier water releases that would increase the reservoir's reserve capacity and its ability to cope with surging flows. Together, these measures would make the dam and reservoir more effective in controlling Iowa River flooding below the dam.

Acknowledgments

I would like to acknowledge the assistance of Lon Drake, emeritus professor of geology at the University of Iowa; David Eash, U.S. Geological Survey, Iowa City; and John Castle, U.S. Army Corps of Engineers, Coralville Dam.

Note

1. Mean sea level (msl): the elevation of the reservoir surface above average sea level. All elevations in this chapter are elevations above msl.

References Cited

Linder, M. 2008. "Give land back to the Iowa River." Iowa City Press-Citizen. July 12, 2008.

Martin, Dave. 2008. Water Control Engineer, U.S. Army Corps of Engineers, Rock Island District Headquarters. Personal communication, August 7, 2008.

Rahn, P. H. 1975. "Lessons learned from the June 9, 1972 flood in Rapid City, South Dakota." *Bulletin of the Association of Engineering Geologists* 12: 83–97.

U.S. Army Corps of Engineers (Corps). 1984. *Emergency Plan for Coralville Dam and Lake*. Rock Island, Ill.: U.S. Army Corps of Engineers, Rock Island District.

———. 1991-a. *Regulation Manual for Coralville Lake*. Rock Island, Ill.: U.S. Army Corps of Engineers, Rock Island District.

———. 1991-b. *Water Control Plan with Final Supplement Environmental Impact Statement, Coralville Reservoir, Iowa*. Rock Island, Ill.: U.S. Army Corps of Engineers, Rock Island District.

———. 2001. *Water Control Manual, Coralville Lake*. Rock Island, Ill.: U.S. Army Corps of Engineers, Rock Island District.

U.S. Geological Survey (USGS). 2008. Water Resources Division, National Water Information System: Web interface, peak streamflow for Iowa, Iowa River at Iowa City, IA. http://nwis.waterdata.usgs.gov/ia/nwis/peak. Accessed fall 2008.

Weber, I. B. 1985. "The Iowa River: Floods were its dark side." In *Irving Weber's Iowa City, Volume 3*, pp. 255–260. Iowa City: Iowa City Lions Club.

Eugene S. Takle

12 Was Climate Change Involved?

DID CLIMATE CHANGE feed into the severe 2008 floods? This is a provocative question. The science of attributing specific climate features to global climate change is new and controversial, especially when applied to weather extremes at the regional or local scale. It is far more defensible to assert that particular global physical factors, such as volcanoes, fluctuations in solar output, greenhouse gases, or sulfate aerosols, are responsible for changes in mean climate features at the global scale. But when rare and extreme weather events seem to increase in frequency, either locally or regionally, both statisticians and thoughtful lay people begin to wonder if something unusual is going on. They ask not only whether climate change was involved, but also—and more urgently—whether such extreme conditions will be repeated soon or nearby. This question is much more than academic for those who are engaged in rebuilding or making other long-term personal, business, or economic decisions while recovering from a weather-induced disaster.

Although land use and other factors may contribute to modern flood events, both climatic data trends and models of future climate suggest that we must also consider our changing climate in discussions of mitigating future floods.

Several types of precipitation trends are important for flooding: total annual precipitation, seasonal total precipitation, recurrent rain events (i.e., 20-day totals discussed below), frequency of intense rain events, and seasonality and timing of precipitation (i.e., increases in spring may cause more flooding than equivalent increases in late summer; rain on snow-covered surfaces can trigger flooding). Our understanding of climate change is based on two research techniques: first, we can examine precipitation trends over the last 30 years (when human-induced or anthropogenic climate change has presumably become a significant factor), and then compare these short-term trends to those of the past 100 years (when anthropogenic factors were minimal although other long-term trends could have contributed). Second, we can employ global and regional climate models (computer-generated numerical simulations of past and future climate based on the laws of physics), which have demonstrated accuracy for simulating the global climate of the 20th century (IPCC 2007) as well as conditions leading to regional droughts and floods (Anderson et al. 2003).

Using these tools, we see that eastern Iowa has experienced increased pre-cipitation of 1 to 2 inches in spring (April through June) over the last 30 years. This is consistent with increases throughout the central U.S. since about 1976 (Groisman et al. 2005). There also is increased intensity of extreme events in the warm season. Groisman et al. (2005) report a 20 percent increase in the most intense 0.3 percent of precipitation events in the central U.S. over this period. By contrast, there has been a slight decrease in the frequency of light or average precipitation events (CCSP 2008). Records from Cedar Rapids (IEM 2008) show that there were 14 days from 1901 to 1950 that had three or more inches of daily total precipitation. Between 1951 and 2000, this number rose to 23 days. Over the last 113 years, annual precipitation in Cedar Rapids has increased by about 9 inches, from 28 to 37 inches. Increases have come in both the warm season and cool season, with the cool season precipitation currently being about 50 percent higher than a hundred years ago. The Cedar Rapids record agrees with the regional trend of increased precipitation since 1976, but the Cedar Rapids upward trend started much earlier. So although it is hard to argue that this locale's increase in annual total precipitation is due to anthropogenic effects of the last 30 years, models suggest this existing trend will continue. The increase in the number of days with intense precipitation, by contrast, has increased in the latter part of the 20th century, which is consistent with changes attributable to anthropogenic effects.

How unusual was the precipitation event in late May and early June of 2008? In some ways, it was very rare indeed. The 20-day accumulated rainfall for

May 23 through June 12 (plate 3), the period immediately preceding the most extreme flooding, produced an annual exceedance probability (probability of being exceeded in any given year) between 1/200 and 1/1000, with a few points exceeding 1/1000, for the hardest-hit locations in Iowa (NOAA 2008). That is, 20-day rainfall totals of this size are estimated to occur on average, and in the long term, only once during any 200- to 1,000-year period. Said yet another way, there is only a 0.5 percent to 0.1 percent, or even less, chance of such rainfalls in any given year.

Of course, the entire Iowa and Cedar River watershed did not experience such extreme precipitation. In the Cedar River basin, for example, the hardest-hit regions centered on the Shell Rock and upper Cedar Rivers; elsewhere the annual exceedance probability ranged from 1/200 to 1/50 or less. Also, these precipitation amounts alone do not explain the 2008 floods; because rivers in Iowa respond to rainfall in a matter of days to a week, it is also necessary to look at rainfall accumulations over shorter time periods to understand the factors contributing to extreme flooding (see chapter 2).

Projections of future climate by global models for the central third of the U.S. (IPCC 2007) generally show trends of increasing precipitation in the north and decreases in the south. However, because global climate models have coarse resolution, we are unable to say with high certainty where the dividing line between increasing and decreasing precipitation might be. Regions to the west of the Midwest likely will become drier, and regions to the east may become more moist. Also there may be a shift in seasonality, with more precipitation coming in the first half of the year and less in the second half. And finally, the frequency of extreme high-intensity events likely will increase. These projected trends are generally consistent with regional (although not necessarily station-by-station) weather trends over the last 30 years (IPCC 2007).

The shift toward more precipitation in late winter, spring, and early summer suggests a tendency toward more spring flood events and increased soil erosion, at least under current agricultural land-management practices. The tendency toward more of the total rain coming in heavy-rain events (true for observed events and projected for future events) means that rainy days likely will be separated by longer periods without rain. This suggests large fluctuations in streamflow and could well mean longer periods of drought.

What might all of these climate changes mean for future midwestern streamflow? Could streamflow increase, and could this imply larger or more frequent floods? Our Iowa State University streamflow modeling research group has conducted numerical simulations relating streamflow to precipitation for the

Upper Mississippi River Basin (figure 12-1) (Jha et al. 2004). When precipitation is low, nearly all of the rain soaks into the ground and almost none runs off immediately into streams. But as precipitation gets heavier, a larger fraction runs off directly into streams. Our model simulation of a future climate (labeled Future in figure 12-1) shows two distinct differences between current and future precipitation events: first, the line for future climate is displaced to the right of the present climate line. This indicates a higher mean annual precipitation in

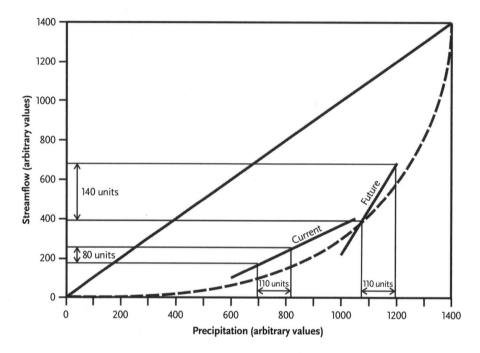

FIGURE 12-1 A graphical depiction of the basic changes in the relationship between stream-flow and precipitation in the Upper Mississippi River Basin (UMRB). The Current climate line (based on measurements of annual precipitation and streamflow) shows that at the present time, when precipitation in the UMRB increases by 110 units, streamflow goes up gradually (by 80 units). In future years (see Future line), the model predicts that the UMRB will have higher mean annual precipitation. In addition, the steeper slope of the Future line indicates that more precipitation will come in high-intensity events. This in turn will result in proportionally more streamflow (an increase of 75 percent, to 140 units, for 110 units of precipitation) and increased potential for flooding. The solid diagonal line gives the hypothetical condition where all precipitation ends up as streamflow, so only points below this line are physically possible. The dashed curve gives the typical response of a watershed to increasing precipitation (essentially no streamflow is created from very light rains, and essentially all precipitation ends up as streamflow for very heavy rains). *Illustration by the author, based on data from Jha et al. 2004.*

coming years. Second, the increased steepness in slope of the future climate line is a subtle but important result indicating that more of the annual total precipitation will come in high-intensity events, resulting in proportionally more streamflow when precipitation increases. Note that comparable increases in rainfall (110 units) in the current and future climates will lead to a 75 percent increase in streamflow (from 80 to 140 units) in the future scenario.

Now, let's return to this chapter's original question: Might climate change have contributed to the 2008 floods? No one can say for certain that this particular climate event was spawned by climate change. However, the statistically documented weather trends and global climate models make a compelling case that climate change may have played a role. A trend toward higher likelihood of extreme precipitation events is observed in the last 50 years for Cedar Rapids and is projected to continue in the future climate. The trend toward larger annual total precipitation in Cedar Rapids, while consistent with projected future trends for this region, began before the effect of greenhouse gases were a contributing factor. Models project at least modest annual total precipitation increases in the future climate for this region. Are these climate models valid? For several reasons, we have confidence that they are. Model simulations of global-averaged climate conditions over the twentieth century are highly correlated with observed conditions. This correlation gives confidence in model skill at the global scale. Correlations with regional trends over this time are much lower but still significant. Rapid warming in the Arctic, a trait that was predicted by all global models and is well documented by observations, is a clear sign of models' regional accuracy at high latitudes. Observed large-scale drying in the western U.S. and southern Europe also is well represented by global models, another sign of their regional skill. Trends in precipitation observed in our region and projected into the future by global models are also consistent and admit a higher likelihood of periods with exceedance probabilities such as those shown in plate 3.

For the reasons I have given above, it is likely that the dice have been loaded toward a higher probability of extreme flood events, with more occurring now than 30 years ago, and with even higher-frequency precipitation conditions leading to such floods in the future. This knowledge and the fact that two major flooding events have occurred in Iowa over the past 15 years should provide strong motivation for public and private decision-makers to seek appropriate ways to enhance Iowa's flood resilience. The scientific community has much to offer in evaluating alternative strategies for developing flood resilience, and decision-makers are strongly encouraged to engage scientists in this effort.

References Cited

Anderson, C. J., et al. 2003. "Hydrological processes in regional climate model simulations of the central United States flood of June–July 1993." *Journal of Hydrometeorology* 4: 584–598.

Climate Change Science Program (CCSP). 2008. *Weather and Climate Extremes in a Changing Climate. Regions of Focus: North America, Hawaii, Caribbean, and U.S. Pacific Islands.* A Report by the U. S. Climate Change Science Program and the Subcommittee on Global Change Research, ed. T. R. Karl, G. A. Meehl, C. D. Miller et al. Washington, D.C.: Department of Commerce, NOAA's National Climatic Data Center.

Groisman, P. Y., R. W. Knight, D. R. Easterling, T. R. Karl, G. C. Hegerl, and V. N. Razuvaev. 2005. "Trends in intense precipitation in the climate record." *Journal of Climate* 18: 1326–1350.

Iowa Environmental Mesonet (IEM). 2008. "IEM Climodat Reports: Cedar Rapids-1." Ames, Iowa: Iowa State University Department of Agronomy. http://mesonet.agron.iastate.edu/, accessed August 2008.

Intergovernmental Panel on Climate Change (IPCC). 2007. *Climate Change 2007: The Physical Science Basis. Contribution of Working Group I to the Fourth Assessment Report of the Intergovernmental Panel on Climate Change*, ed. S. Solomon, D. Qin, M. Manning et al. Cambridge, United Kingdom, and New York: Cambridge University Press.

Jha, M., Z. Pan, E. S. Takle, and R. Gu. 2004. "Impact of climate change on stream flow in the Upper Mississippi River Basin: A regional climate model perspective." *Journal of Geophysical Research* 109, D09105, doi:10.1029/2003JD003686.

National Oceanic and Atmospheric Administration (NOAA). 2008. "Annual exceedance probabilities for June flooding in the Upper Midwest." National Weather Service, Office of Hydrologic Development, Hydrometeorological Design Studies Center. Private communication, Geoffrey Bonnin, August 2008.

Flood Damages, Flood Costs, Flood Benefits

FROM TIME IMMEMORIAL, humans have been drawn to rivers and settled along their banks. Rivers provided the water needed by people and their livestock and also attracted game for the taking. Fertile alluvial soils often abounded here. People used the rivers as transportation corridors and as a source of power for mills that ground grains and cut lumber.

Iowa's rivers attracted people and served all of these functions during the settlement era. Good riverside mill sites were among the most highly prized lands, farm goods were easily floated downriver to St. Louis, and timber for structures and firewood was abundant along water corridors. Communities sprouted up where water was easily accessible, even though cattle were sometimes trapped or lost in bottomland sloughs, malaria was prevalent, and sporadic floods could reclaim control of the floodplain in a day. In spite of such dangers, Iowa was first settled along its rivers. Look at a state map: virtually every city of any size is located on a river or stream.

Today, the character and beauty of our communities continue to be enhanced by their rivers. Housing sites near water, even on floodplains, still appeal to many people, despite the inherent danger of living there. Why? Because extreme floods are rare, and our lives and our memories are short. And because at some level, the risks of living in floodplains—if they are understood at all—are judged as being outweighed by the benefits.

For all these reasons, people too often find themselves in the wrong place at the wrong time: seeking safety on a rooftop or evacuating flooded homes by rowboat or canoe. Then the costs and destruction of floods take precedence over a river's amenities, and the river's natural behavior is condemned. Yet if we are honest with ourselves, we see that flood costs are defined by us rather than by the river. We have chosen to site our homes, our cities, and our lives in harm's way; we have chosen to embrace the floodplain and its risks.

If we stand back from flooded cityscapes and take a broader view, it becomes evident that floods, over long periods of time, have produced many benefits. They have shaped our landscape. With their sediment deposits, they have produced and continued to nurture some of our richest agricultural soils. And they have interacted with riverside plant and animal communities to produce sustainable mixtures of native species that are not just flood-tolerant but flood-dependent. Several types of riverside natural communities and native species require periodic scouring by floods in order to survive, and their survival safeguards floodplain integrity and biodiversity.

This section considers the complex mix of flood costs, damages, and benefits. Much of the section looks at the damages wrought to the floodplain portions of Cedar Rapids and Iowa City/Coralville. But agricultural lands are also considered, as are the potential benefits to archaeological sites and natural areas. Joe Alan Artz and Lynn M. Alex begin by looking at the high concentrations of archaeological sites found along waterways, sites that show rivers have been important to Iowa's inhabitants for over 10,000 years (chapter 13). About a third of Iowa's known precontact Native American sites are in areas subject to flooding. Examples show that floods can erode or wash away some archaeological sites, but they also cover other sites with protective layers of sediment.

The next two chapters continue with straightforward descriptions of 2008 flood effects on modern humans. Chapter 14, composed of statistics from multiple sources, lists specific damages and costs reported for selected Johnson and Linn County sites. In chapter 15, Daniel Otto looks at Iowa's financial costs, explaining that the floods were one of the nation's largest natural disasters in 2008. Describing some of the difficulties encountered when calculating total flood-related costs and enumerating several categories of losses, the author comes up with an Iowa bill of $3.506 billion—considerably larger than Iowa's 1993 flood costs.

While cities were hit hard, agricultural land suffered at least as much, as Richard Cruse, Hillary Olson, and John M. Laflen describe in chapter 16. Because the heavy rains fell when crops were small and the soils had little protective cover, erosion rates were stupendous—in some townships reaching 50 tons per acre in a single day, ten times the annual acceptable average. The authors describe the erosion process as well as flood damages to agricultural terraces and waterways.

The next two chapters address the content of floodwaters and air quality hazards. In chapter 17, Dana Kolpin and Keri Hornbuckle explain that heavy rains and resulting floodwaters had the opportunity to pick up a tremendous variety of natural contaminants (e.g., animal manure, waste from wastewater treatment plants, and bacteria) and synthetic chemicals (commercial fertilizers and pesticides such as atrazine and acetochlor, cleaning supplies, industrial chemicals such as PCBs, etc). These came from both agricultural and urban sources, the latter including homes and industry. The effects of this complex "chemical cocktail" on humans and wildlife are worrisome but unknown; more study of these contaminants is badly needed. In chapter 18, Peter S. Thorne considers air quality hazards, pointing out that both chemical air contaminants (carbon monoxide, asbestos, and pesticides) and organic contaminants (mold spores, microorganisms, and endotoxins) are created by floods or released during flood cleanup operations.

The section ends with John Pearson's discussion of flood effects on natural communities (chapter 19), which once again reflects the benefits as well as damages invoked by flooding. Floodplain communities are often renewed and restored by floods, which can fill wetlands, scour out new seedbeds for certain floodplain plants, and create habitat for water organisms. However, the effects of flooding need to be considered for each species involved, as does the fragmentation of modern native communities. The boundary between benefit and cost is not always clearcut. As with most flood questions, more study is needed.

These chapters, with their tremendous diversity of subject matter and mixture of positive and negative flood effects, point out the tremendous complexity of issues associated with floods. This complexity continues into the next and last section of the book, which describes ways of moving forward and forging a healthier relationship with our landscape and the floods that are sure to come. ≋

Joe Alan Artz
Lynn M. Alex

13 Flood Effects on Archaeological Sites

PEOPLE HAVE LIVED in Linn and Johnson Counties for over 10 millennia. For most of that time, the only records of their presence are the archaeological remains of villages and campsites where they lived and worked, and the cemeteries and mounds where they buried their dead. Although artifacts like arrowheads and pottery sherds can be found virtually anywhere, most such finds are located within a mile or two of streams. This concentration indicates that rivers and their floodplains were important to ancient people, providing water, food, shelter, tillable soil, and arteries of transportation.

The Office of the State Archaeologist (OSA), a research unit of the University of Iowa, is charged by state law with maintaining records of Iowa's archaeological sites. The OSA currently has information on over 17,000 precontact sites, those occupied by Native Americans prior to the arrival of European Americans. Of these, about 5,800 are located on floodplains or in areas that have been subject to flooding at some time during the past 10,000 years (figure 13-1).

Rivers influenced not only where ancient people chose to live, but also what subsequently became of their sites. Floods can damage or destroy archaeological sites, for example, by cutting away the remains of an ancient village. Or, floods can contribute to their preservation, by burying sites under protective layers of silt. It will take months to assess the damage done to Iowa's archaeological resources by the floods of 2008, but certain kinds of effects can be anticipated. Fishel and Van Nest (1994) visited 76 of the state's most important streamside archaeological sites after the severe floods of 1993. Eighteen were covered with fresh deposits of sediment, 26 had eroded, and 9 experienced both erosion and deposition. Surprisingly, 23 sites showed no sign of having been affected by flooding. Most damage was caused by the collapse of stream banks. A stream responds to flooding by widening its channel to contain the increased volume of water. As the banks are eaten back, ancient artifacts buried in the collapsing soil also will fall into the water and be washed away.

On the other hand, freshly cut stream banks may expose never-before-seen archaeological remains. For example, following the 1993 floods, excavation of

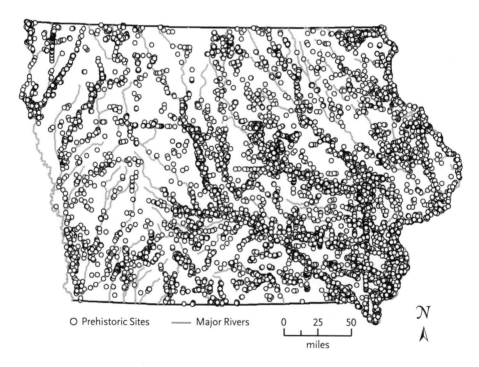

O Prehistoric Sites   —— Major Rivers   0   25   50

miles

N

FIGURE 13-1 Recorded archaeological sites in Iowa. Overlapping circles show areas where multiple sites are documented. Straight lines of circles denote those discovered during modern road surveys. The concentration of sites along waterways illustrates the importance of river and floodplain resources to ancient people. *Illustration by Geospatial Program, University of Iowa, Office of the State Archaeologist.*

flood-exposed deposits at the Dixon site, near Sioux City, revealed new information about the diet, technology, and community organization of people living in western Iowa 700 years ago (Fishel 1999).

Similar results can be anticipated when evaluating the 2008 flood impacts to Johnson and Linn Counties. Over the years, 5 percent (73 square miles) of these two counties have been systematically surveyed by archaeologists and found to contain 881 prehistoric archaeological sites. Of these, 293 are in areas that were inundated by the 2008 floods (figure 13-2).

One site in Johnson County that clearly illustrates the preservative effect of floods is the Edgewater Park site (Whittaker et al. 2007). This 3,800-year-old Native American campsite, located along the Iowa River in Coralville, is buried under four feet of river sediment. Artifacts found here by excavators include hundreds of small chert flakes from the final stages of stone tool making. Two fire hearths yielded charred seeds from plant species that mature during the late summer. The kinds of artifacts and the small size of the site indicate it was occupied for short periods of time by small groups of nomadic people.

Examination of the river sediments that cover the site reveal the likely reason that the site was infrequently used. Over time, the iron minerals in soil that is above the water table tend to oxidize to yellow and red colors. Below the water table, the minerals chemically reduce to grays, blues, and greens. The Edgewater Park site is buried in soil that is predominantly yellowish-brown in color, suggesting that the site was relatively dry, but splotches of gray colors indicate that the water table fluctuated, perhaps seasonally. The picture is one of a small group of people occupying a low but seasonally dry ridge on the floodplain, probably during the late summer when the water table was low. Perhaps, 3,500 years ago, the inhabitants knew better than to try living year round in a place where the water table was often high and floods were probably frequent. In July 2008, as the nearby Marriott Hotel sat surrounded by floodwater, the Edgewater Park archaeological site, although submerged, accumulated a few more millimeters of protective silt.

Another example of a site preserved by flooding is located in the Wickiup Hill Natural Area, north of Cedar Rapids. Excavations turned up small quantities of pottery and stone artifacts from the Late Woodland period, A.D. 600–900. One excavation came across two complete, although crushed, pottery vessels, found less than a foot apart, two feet below the present surface (Shields 1999). The silty soil in which they were buried was deposited in thin layers by floods that spread gently over the floodplain.

The most intensively occupied prehistoric sites at Wickiup Hill are located

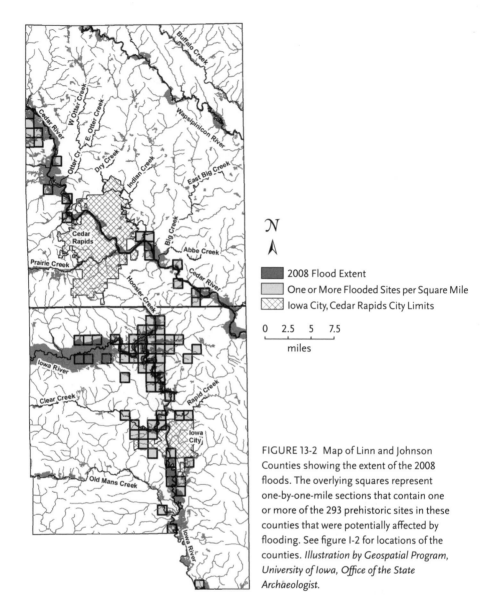

N

⋀

■ 2008 Flood Extent
▢ One or More Flooded Sites per Square Mile
▨ Iowa City, Cedar Rapids City Limits

0   2.5   5   7.5
———————————————
miles

FIGURE 13-2 Map of Linn and Johnson Counties showing the extent of the 2008 floods. The overlying squares represent one-by-one-mile sections that contain one or more of the 293 prehistoric sites in these counties that were potentially affected by flooding. See figure I-2 for locations of the counties. *Illustration by Geospatial Program, University of Iowa, Office of the State Archaeologist.*

on Ice Age sand dunes, 20 to 40 feet above the Cedar River floodplain (Rogers and Green 1995). The sandy, well-drained soils of the dunes were more suitable for habitation than the lower-lying floodplain. The picture at Wickiup Hill is one of ancient people occasionally inhabiting a flood-prone area, but more often choosing higher, better drained land.

A third example is of a site that has been both eroded and preserved by flooding. Woodpecker Cave, perched 10 feet above a small stream draining into Coralville Reservoir, is a rock overhang formed in Devonian limestone (figure 13-3). Such overhangs offered shelter to ancient people throughout eastern Iowa. The cavern provides a sunny, southern exposure with its backside a buffer to prevailing winds. The site is inundated during extreme floods. In anticipation of this threat, Smithsonian Institution archaeologists excavated at the site before the dam was constructed. They discovered two distinct layers

FIGURE 13-3 Woodpecker Cave, Johnson County. People periodically occupied the area beneath this protective overhang for at least 3,000 years. In 2008, floodwaters filled the site, reaching to the top of its rocky ceiling. *Photograph by Lynn M. Alex.*

of human habitation within the cave (Caldwell 1961). The topmost layer contained a rich assortment of stone tools, pottery, and discarded animal bones, demonstrating intermittent occupation by people for over 3,000 years. In the lower level, ash-strewn earth, discarded bone, and stone flakes confirmed even earlier occupation.

In 2008, water from Coralville Reservoir rose well above the 17-foot-high overhang, completely submerging the site. The receding water spread a thin layer of silt over the shelter floor but eroded some of the rocky slope in front of the cave, encroaching on archaeological deposits that may remain deeply buried at the site.

Floods also can affect archaeological sites by changing a river's course. Stream meanders, like those of the Iowa River, migrate across floodplains in response to the flow of water. The stream erodes its outer banks and deposits sand and mud on the inside of its bends. During floods, as the water velocity increases, bank erosion is accelerated. Then, as the river level rises, the water cuts across low-lying areas on the inside of meander bends. A flood of sufficient magnitude will erode a new channel across the neck, making an oxbow lake of the former bend.

This process is known to have affected archaeological sites on the Iowa River, just south of Iowa City. Here, in the 1830s, about 1,600 Meskwaki Indians lived in two villages. In 1835, John Gilbert, an agent for the American Fur Trade Company, established a trading post near the southern village. At the time, as shown on early maps, both villages and the trading post were situated on the banks of the Iowa River, on the outside of meander bends (Peterson 1997). Sometime between 1870 and 1900, the river cut a new channel across the meander at the southern village, leaving it on the shore of an oxbow lake. Sometime between 1917 and 1930, another meander cutoff left the site of Gilbert's trading post about 2,000 feet east of the river. During this same period, the northern village was destroyed by the northeasterly migration of a river meander. Peterson (1997) relied on maps of former river channels to successfully relocate Gilbert's post. Her archaeological investigation revealed that, in the years following the post's abandonment, flood deposits had accumulated over it, preserving the site and the story it tells about trading post life.

In closing, as great a tragedy as Iowa's recent flooding was to those who lost their homes and businesses, the archaeological sites found buried by river sediment in the stream valleys of Johnson and Linn Counties remind us that the experience of 2008 was not unique. When the Iowa River rose from its banks 3,500 years ago and covered the campsite at Edgewater Park, it was undoubtedly responding to heavy precipitation to the north and west in the Iowa River watershed. Two millennia later, an overflowing Cedar River, swollen from heavy rain or perhaps upstream snowmelt, spread silt across two clay pots left in place at a prehistoric campsite at Wickiup Hill. Almost within memory, the meandering Iowa River, perhaps empowered by floods, both destroyed and afforded protection to an important piece of Iowa's territorial history. While we can't know whether Iowa's earliest residents ever ran for high ground or watched helplessly as their homes disappeared beneath rising floodwaters, the archaeological record clearly demonstrates that the floods of the past had consequences to people and the places where they chose to live and work.

References Cited

Caldwell, W. W. 1961. *Archaeological Investigations of the Coralville Reservoir, Iowa*. River Basin Surveys Papers No. 22, Bulletin 179, pp. 79–148. Washington, D.C.: Bureau of American Ethnology, Smithsonian Institution.

Fishel, R. L. (editor). 1999. *Bison Hunters of the Western Prairies: Archaeological Investigations at the Dixon Site (13WD8), Woodbury County, Iowa*. Report 21. Iowa City: Office of the State Archaeologist, University of Iowa.

Fishel, R. L., and J. Van Nest. 1994. *Preliminary Flood Impact Assessment of Selected Iowa Archaeological Sites Damaged in 1993*. Contract Completion Report 412. Iowa City: Office of the State Archaeologist, University of Iowa.

Peterson, C. L. 1997. *Sand Road Heritage Corridor, Johnson County, Iowa: Archaeology and History of Indian and Pioneer Settlement*. Contract Completion Report 492. Iowa City: Office of the State Archaeologist, University of Iowa.

Rogers, L. D., and W. Green. 1995. *Wickiup Hill Natural Area Archaeological Survey*. Research Papers Vol. 20, No. 3. Iowa City: Office of the State Archaeologist, University of Iowa.

Shields, W. L. 1999. *A Phase I Archaeological Survey of Primary Roads Project NHS-30-6(62)–19–06, Benton and Linn Counties*. Project Completion Report Vol. 22, No. 20. Iowa City: Office of the State Archaeologist, University of Iowa.

Whittaker, W. E., M. T. Dunne, J. A. Artz, S. E. Horgen, and M. L. Anderson. 2007. "Edgewater Park: A late archaic campsite along the Iowa River." *Midcontinental Journal of Archaeology* 32: 5–46.

Cornelia F. Mutel

14 Flood Effects on Modern Communities

THE 2008 FLOODS can be measured, described, and studied, their causes sought, and certain types of their effects measured. But these efforts give little vision of the magnitude of flood impact on modern residents: displaced families and lost homes, family treasures, and jobs. Many aspects of these damages will never be recorded. However, they can be sensed to a small degree in statistics.

Of course, no one was tallying data on losses while the waters raged. That job was left to city and county employees who themselves were often displaced from their offices, workers who were scrambling to construct a future even as they recorded the past. Here is a listing of raw damage statistics for Johnson and Linn Counties and entities within these counties. The listing includes data available at the time; thus categories of loss are not consistent from one county or city to the next. In addition, some of the data remain preliminary. They were provided by government administrators on the dates indicated (see page 137), but numbers may change as more statistics become available. Additional information on financial costs of flooding for the entire state (with a few references specifically to Johnson and Linn Counties) is included in chapter 15. This listing does not do service to the amount of loss in rural areas and farms and in small towns (figure 14-1).

FIGURE 14-1 Although statistics in this chapter focus on urban losses from the 2008 floods, Iowa's small towns and rural areas also were intensely affected. The losses and costs of destruction to farmlands and human structures there were tremendous (see chapters 15 and 16). *Photograph by Lindsey Walters,* Daily Iowan.

Cedar Rapids

Extent of flooding

- 9.2 square miles (about 14 percent of city)
- At Cedar Rapids streamgage: maximum discharge 140,000 cubic feet per second (cfs) (average discharge 3,807 cfs), on June 13; high water, 31.12 feet (flood stage, 12 feet); 13 days above flood stage—June 8–21

Persons affected (estimated)

- 18,600 total
- 1,834 Cedar Rapids Community School District students resided in flooded areas

Structures damaged

- 7,198 parcels, including around 5,400 residential, 1,050 commercial, and 84 industrial properties
- Eight cultural assets, including the Islamic Center of Cedar Rapids,

National Czech & Slovak Museum & Library, African American Museum of Iowa, and Cedar Rapids Museum of Art
- Eight Cedar Rapids Community School District facilities

Costs
- $33.5 million, estimated Cedar Rapids Community School District damage

Additional impacts
- Water supply: 3 out of 4 city collector wells, 46 vertical wells disabled by flood
- 72,428 tons of debris collected and removed to landfills (as of 9/24/08) (figure 14-2)

FIGURE 14-2 Flood recovery begins with the disposal of staggering amounts of debris. Nearly every household item soaked by floodwater is discarded. Most residents helped by sorting and separately stacking household hazardous waste, electronics, and appliances for proper disposal. *Photograph by Richard A. Fosse, city of Iowa City.*

Linn County
Structures damaged
- More than 250 in unincorporated county
- 10 county buildings, including the Administrative Office Building, County Correctional Center, Sheriff's Office, County Courthouse, Youth Shelter
- 12 arts and culture organizations and 37 human service organizations; rebuilding costs to exceed $50 million

Costs
- $64.807 million, damage estimate for Linn County government (including lost revenue)

Iowa City
Extent of flooding
- 1,600 acres
- At Iowa City streamgage: maximum discharge 41,100 cfs (average discharge 2,300 cfs), on June 15; high water, 31.53 feet (flood stage, 22 feet); 32 days above flood stage (June 5–July 7)

Persons affected
- 1,531 housing units evacuated

Structures damaged
- 251 housing units, 52 commercial properties (figure 14-3)
- 3 of 5 bridges closed during flood; remaining 2 bridges within 5 inches of flooding

Costs
- $6.47 million, damages to public land and structures, response efforts

Sandbagging efforts
- 26,040,020 pounds of sand hauled, more than 1.5 million sandbags installed

Additional impacts
- 6 of 11 municipal wells out of service
- 8,656 phone calls to Flood Information Center

Coralville
Extent of flooding
- 273 acres
- Flooding began June 5; Highway 6 remained under water until June 28, Edgewater Drive through July 6

FIGURE 14-3 The Idyllwild neighborhood was one of a number of Iowa City residential areas affected by the flood. This relatively new neighborhood (construction started around 1990) demonstrates the limits of protection from extreme events afforded by current floodplain development regulations (see chapter 20). *Photograph by Ron Knoche, city of Iowa City.*

- Water 8 feet deep in street centerlines, 6 feet in Transit Maintenance Facility

Persons affected

- 310 housing units evacuated

Structures damaged

- About 200 businesses, 400 households

Sandbagging efforts

- 19 million pounds of sand used, 1 million sandbags placed

The University of Iowa

Persons affected

- Faculty, staff, and students, and their teaching, research, and artistic activities evacuated from offices, laboratories, classrooms, studios, practice rooms, rehearsal and performance space. Displaced for weeks or months, some potentially for years.

Structures damaged

- 21 buildings evacuated; 22 major structures impacted by floodwaters including the entire School of Music and most of the School of Art and Art History complex
- 2.4 million square feet of campus building space closed or damaged by the flood
- 25 percent of general assignment classrooms located in flooded buildings and temporarily closed. Also research laboratories and equipment, practice and performance rooms, student housing, graduate student apartments, softball and track and field complexes, service buildings.

Costs

- $232 million in damages, plus expenses of business interruption, leased replacement space, long-term protection of recovered buildings, and potential replacement of flood-damaged facilities (for total anticipated cost of flooding, see chapter 4, footnote 1)

Sandbagging efforts

- More than 5,000 volunteers

Additional impacts

- Majority of university offices and all summer classes and sports camps suspended, June 13–June 23
- Main Power Plant shut down June 14, officially reactivated at full function on October 28; 42 miles of cabling and wiring and 3 miles of pipe insulation replaced
- 1.34 miles (of 5.9 miles total) of utility tunnels beneath campus flooded; 4.83 miles of steam pipe insulation replaced in these tunnels

Johnson County, outside municipal limits

Structures damaged

- About 110, 44 substantially
- Collapse of historic Sutliff Bridge (constructed 1898) over Cedar River

Costs

- Approximately $1.15 million in flood-related labor and equipment, bridge checks, debris cleanup; FEMA estimate of total damage, $2.1 million (as of 11/1/08)

Acknowledgments

Thanks to the following contributors for statistical and other information: Cedar Rapids—Cassie Willis, communications liaison, Cedar Rapids, 10/28/08, confirmed 2/28/09, excluding school district information which was provided by Marcia Hughes, community relations supervisor, Cedar Rapids Community Schools, 4/6/09; Linn County—Joi Bergman, communications director, Linn County, Les Beck, planning and development, Linn County, and Terry Bergen, director of marketing and communications, United Way, 3/15/09; Iowa City—Richard A. Fosse, director of public works, City of Iowa City, 11/28/08, confirmed 2/24/09; Coralville—Ellen Habel, assistant city administrator, Coralville, 9/9/08, confirmed 2/24/09; the University of Iowa—Rodney Lehnertz, director of planning, design, and construction, University of Iowa, 12/15/08, confirmed 2/24/09; Johnson County outside municipal limits: Richard J. Dvorak, planning and zoning administrator, Johnson County, 11/5/08, confirmed 3/3/09.

Daniel Otto

15 Economic Losses from the Floods

FLOODING IN IOWA was one of the largest natural disasters in the U.S. in 2008. While from a national accounting perspective (Mattoon 2008) the 1993 floods in the upper Midwest were larger in scope and damages, the 2008 flooding appears to have caused more damage and losses for Iowa communities and individuals. The most severe flood damages occurred in communities along the Cedar and Iowa Rivers in eastern Iowa (RIO 2009). This chapter examines the financial impact of Iowa's 2008 flooding, with some comparisons to economic losses in the 1993 floods. The chapter was written before reliable statistics on the year's crop production were available and when losses of structures and infrastructure and economic activity were still preliminary; thus figures in this chapter are subject to revision. Even several months after the end of the flooding, information on the magnitude of the economic losses remained sketchy, with the quality of information varying for different categories of property.

Although damages from flooding are immediately obvious, developing a comprehensive and reliable estimate of economic losses is very difficult. Nevertheless, interest in estimating the magnitude of economic losses begins almost immediately as public officials prepare requests for disaster assistance and begin planning for recovery. The National Research Council of the National Academy of Sciences has produced guidelines for assessing economic losses from natural disasters (NRC 1999). That report proved helpful in assessing economic losses from Iowa's 2008 floods.

Natural disasters like floods and tornados affect the economy directly through the destruction of private homes, businesses, and public infrastructure. This property destruction damages the economic capacity of a region and disrupts normal economic activity. While spending on rebuilding becomes economic gain for the construction and building supply sector, there is loss of future growth potential because public and private resources are diverted away from other productive uses and into flood recovery. However, because most of the recovery costs come from federal sources, reconstruction spending should have a stimulative effect on state and regional economies. As of January 2009, $1.4 billion of disaster relief funds, mainly from federal sources, had been earmarked for Iowa flood recovery (RIO 2009).

In 1993, the majority of flood-related losses were agricultural (Otto and Lipsman 1993). That year, wet weather and flooding occurred into July, too late for crops to be replanted. Because of sizeable U.S. crop inventories that year, revenues for flood-affected farmers suffered because weather-related yield reductions were not offset by a significant increase in crop prices.

In contrast, the 2008 flooding in early June allowed many of the crop wash-outs to be replanted. The favorable weather through much of the remaining growing season brought statewide yields back to average expected state levels, according to September and October crop reports (USDA 2008).

To estimate economic effects of the adverse 2008 weather on the agricultural sector, we used projected agricultural statistics to compare likely production of corn and soybeans (Iowa's dominant crops) after the flooding to expected production under normal conditions. On average, about 2.5 percent of planted corn acreage each year is not harvested, and about 0.5 percent of planted bean acreage is not harvested. In fall 2008, the projected unharvested percentages exceeded these long-term averages, reaching 5.8 percent for corn and 1.1 percent for beans. The wet weather also lowered expected yields to slightly below the 28-year trend (from 169.5 to 168 corn bushels per acre, and from 49.25 to 47 soybean bushels per acre). These differentials in projected harvested acres and yields produced 2008 loss estimates of $539.7 million for Iowa corn producers, and $296.7 million for Iowa soybean producers (Hart 2008).

These estimated losses represent statewide aggregates. However, crop losses were concentrated in the eastern half of the state and in lowlands along flooded streams and rivers. In contrast to the floods of 1993, carry-over crop inventories were lower in 2007–2008 so that production reductions resulted in price increases that helped offset some of the total agricultural losses. These price increases were very beneficial for farmers in the western half of the state, who were relatively unaffected by flooding. High crop prices for the season meant that statewide estimates of farmer revenues for 2008 did not suffer as dramatically as they did in 1993. In addition, crop producers often had several safety nets in place, such as yield and revenue insurance and disaster payments paid out on a farm-by-farm basis, which helped offset 2008 flood losses for individual farmers.

The second major category of flood losses, those describing several major categories of private and public physical properties, is summarized in table 15-1. The major source of information for these losses was the September report of

TABLE 15-1 Estimated Economic Losses Associated with the 2008 Floods in Iowa as of September 2008. Source: RIAC 2008.

	$ Million
Structures	
Housing	$946
Rental	$90
Private	$856
Public buildings	$380
Infrastructure	
Roads and bridges	$125
Railroads	$68-83
Utilities	$408
Other transportation	$53
Education facilities	
K-12	$63.7
Regents	$232.1
Other	$1.6
Cultural and historic landmarks	$284.5
Agriculture/environment	
Crop agriculture	$836.4
Environmental damage	$75
DNR facilities	$18
TOTAL	$3,506.3

the Rebuild Iowa Office (RIAC 2008). The report provides a summary of flood damages reported through a variety of state and federal agencies.

Damages to the single family and rental housing stock were estimated to be nearly $1 billion. Dealing with housing losses is likely to present ongoing difficulties because many of the flooded homes were modestly valued smaller and older units, and the cost of providing replacement housing will be greater than the pre-flood value of many of the homes (RIAC 2008).

Public infrastructure damages were fairly extensive. The Iowa Department of Transportation estimated road and bridge damages at $125 million and railroad damages at $68 to $83 million. Other transportation assets had losses of about $53 million. Damage to electric, water, and other public utilities were estimated at $408 million. Other public buildings, including courthouses, libraries, and government offices, suffered $380 million of losses, according to Federal Emergency Management Agency (FEMA) estimates (RIAC 2008).

Educational facilities, especially properties at the University of Iowa, sustained significant damages. Losses at the university were placed at $232.1 million. Other K–12 facilities throughout the state sustained $63.7 million of damages. Those of other cultural facilities such as museums and the Czech Village properties in Cedar Rapids were estimated as $284.5 million (RIAC 2008).

In addition to the agricultural losses estimated for Iowa crop producers, other natural resource-related infrastructure incurred damages. Damages to environmental resources such as waterways and erosion control infrastructures totaled $75 million of estimated losses. Iowa Department of Natural Resources–operated facilities reported losses of about $18 million. In total, this set of identified losses to physical property adds up to an estimated $3.5 billion, including crop losses (RIAC 2008).

Financial losses from flooding also include indirect categories, primarily lost economic activity such as lost commercial sales, temporary unemployment, and disrupted businesses. The Iowa Department of Revenue sales tax data for the affected period provide an assessment of the flood's impact on this portion of the economy (Lipsman 2008). This preliminary flood impact analysis focuses on a detailed set of retail and service businesses such as construction contractors, traditional retailers, and restaurant, bar, and lodging establishments. These business types accounted for between 64 percent and 66 percent of total taxable sales during the two quarters analyzed. The analysis focuses on changes in taxable sales between the June 2008 (April–June 2008) and September 2008 (July–September 2008) quarters.

The percentage change in taxable sales between the June 2008 and September

2008 quarters for seven general business classifications is presented in figure 15-1 for the state of Iowa, Linn County, Johnson County, and the state exclusive of these two counties. In the intensely flood-affected Johnson and Linn Counties, contracting services (electrical, concrete, roofing, siding, painting and wall covering) saw significant increases in sales of products and services related to flood recovery. Between the June 2008 quarter and the September 2008 quarter, taxable sales by building contractors of all types increased by 23.4 percent statewide, by 28.6 percent in Linn County, by 42.3 percent in Johnson County, and by 21.8 percent in the other 97 counties of Iowa (Lipsman 2008).

Significant sales increases also were observed in the furniture and appliances category as apparently families and businesses were replacing damaged or destroyed property. These sales increased 27.1 percent in Linn County and 10.9 percent in Johnson County compared to 5.6 percent in the rest of the state. Also, while general merchandise and clothing sales statewide decreased during

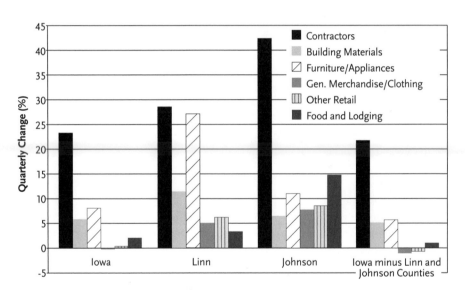

FIGURE 15-1 Percent change in flood-related Iowa taxable sales, June to September 2008. Graph compares pre-flood data (Quarter 2, April–June 2008) to post-flood data (Quarter 3, July–September 2008), and reflects the percent change for different regions and sales categories. Note that the counties most affected by flooding in June 2008 (Linn and Johnson Counties) had significant increases in sales of contracting services, building materials, and general merchandise relative to averages elsewhere in Iowa. These sales categories are most heavily involved in flood recovery efforts. Spending in these categories also provided a general boost to state spending totals on these categories over the previous year. *Illustration by the author, based on preliminary tax data in Lipsman 2008.*

the quarter following the floods, taxable sales by these businesses increased by 4 percent in Linn County and by 7.7 percent in Johnson County (Lipsman 2008). These data suggest that the flooding, which was concentrated in these two counties, did indeed trigger increased spending to repair or replace damaged properties and possessions.

In conclusion, the Iowa floods of 2008 were a very destructive natural disaster with total property losses far exceeding the in-state damages occurring during the 1993 floods. Current estimates place total damages to private property and public infrastructure at over $3 billion, with the highest volume of damages in Linn and Johnson Counties. These estimates do not account for lost sales by businesses and industries in the flooded areas. Offsetting these losses has been a significant inflow of federal disaster relief monies. As of January 2009, an estimated $1.4 billion of federal dollars had been earmarked for disaster relief in Iowa. The combination of these external dollars and the spending to rebuild and replace damaged properties appears to have had a short-run stimulative effect on the Iowa economy, as retail sales in the most severely affected counties were significantly above state averages in these reconstruction-related categories.

Acknowledgments
The author would like to thank Michael Lipsman, manager, Tax Research and Program Analysis Section, Iowa Department of Revenue, for providing the analysis and figures on retail sales performance in the flood counties and for his careful review and comments on drafts of this chapter.

References Cited
Hart, C. 2008. "Estimated value of crop losses from 2008 spring flooding." Worksheet and analysis, August 2008. Ames, Iowa: Department of Economics, Iowa State University.

Lipsman, M. 2008. *Analysis of Flood-Related Retail Spending in Linn and Johnson Counties*, confidential report, 11/2008. Des Moines: Tax Research and Program Analysis Section, Iowa Department of Revenue.

Mattoon, R. July 10, 2008. "Assessing the Midwest floods of 2008 (and 1993)." Chicago: Federal Reserve Bank of Chicago. http://midwest.chicagofedblogs.org/archives/2008/07/mattoon_flood_b.html.

National Research Council (NRC). 1999. *The Impacts of Natural Disasters: A Framework for Loss Estimation.* Committee on Assessing the Costs of Natural Disasters, Commission on Geosciences, Environment, and Resources. Washington, D.C.: National Academy Press.

Otto, D., and M. Lipsman.1993. "Economic impacts of the 1993 Iowa floods." Department of Economics report, Ames, Iowa: Iowa State University.

Rebuild Iowa Office (RIO). 2009. "Rebuild Iowa Office details more than $1.4 billion in disaster assistance," press release, January 2, 2009. Des Moines: Rebuild Iowa Office. http://www.rio.iowa.gov/news/releases/2009/weekly_update/010209_weekly_update.html.

Rebuild Iowa Advisory Commission (RIAC). 2008. *45-Day Report to Governor Chet Culver*. Des Moines: Rebuild Iowa Advisory Commission. http://www.rio.iowa.gov/resources/reports/riac_45-day_report_09-2008.pdf.

U.S. Department of Agriculture (USDA). 2008. "USDA corrects October crop acreage estimates," news release 0278.08, 10/28/2008. Washington, D.C.: USDA newsroom. http://www.usda.gov/wps/portal/!ut/p/_s.7_0_A/7_0_1OB?contentid only=true&contentid=2008/10/0278.xml.

Richard Cruse
Hillary Olson
John M. Laflen

16 How Did the Floods Affect Farmland?

Iowa's FARMLAND IS defined by its soils, which are among the most fertile, productive, and deepest in the world. For centuries these soils were protected from damage by intense rainfall by lush perennial prairie cover, wetlands, and forests. Today our highly productive soils produce food, feed, fiber, and fuel for the world, and about two-thirds of Iowa is covered with annual row crops. Row crops are very productive, but compared to perennial vegetation, they offer soils much less protection against devastating storms. In a typical year, scientists have estimated that the power of impinging raindrops on one acre of Midwest ground is equivalent to the energy found in 20 tons of TNT (Meyer and Mannering 1967).

Growing-season floods are typically associated with extreme rainfall, which erodes soil and reshapes cropland and its productivity. This chapter focuses on erosion and other degradation of soils by the 2008 rains. We will begin our exploration by considering basic aspects of soil science.

Soil productivity is related to the soil's depth, fertility, organic matter content, structure, texture, and bulk density, among other factors. Topsoil—the upper layer of soil, typically 2 to 8 inches deep—contains the best structure (best aggregates), the highest soil organic matter content, and the most nutrients. Half or more of Iowa's original prairie topsoil (which originally averaged about 16 inches deep on mesic uplands) has been lost since the plowing of native prairie vegetation (Mutel 2008). Topsoil depth is a major determinant of crop yields (figure 16-1) for the two major soil types that dominate Iowa: loess—soil material that was deposited by wind; and glacial till—soil material that was deposited by glaciers (Kazemi et al. 1990).

Rainfall erodes topsoil, especially in cropland. Typically we assume that soil erosion rates of about 5 tons per acre per year are tolerable, meaning that we can maintain crop productivity if we limit erosion rates to this level. Five tons

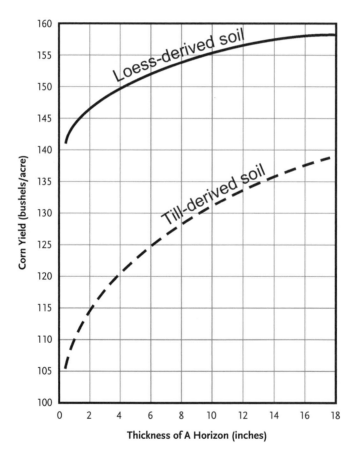

FIGURE 16-1 As topsoil depth (or A horizon) decreases, corn yield decreases for the two main soil types found in Iowa. *Illustration by Richard Cruse, based on Kazemi et al. 1990.*

approximates the thickness of a dime uniformly covering an acre. Greater erosion losses are assumed to negatively affect crop yield. However, some literature suggests that even 5 tons per acre is too much to lose because soils form and are renewed at less than 1 ton per acre per year (Montgomery 2007; Alexander 1988; Owens and Watson 1979).

Soil erosion is the process of detaching or loosening soil particles from the soil surface, transporting them in water flowing over the surface, and redepositing the material either in the field from which it originated or at a location outside the field. Detachment occurs in one of two ways: first, through forces associated with raindrops striking the soil surface, and second, through forces associated with water movement in runoff. In the first process, raindrops strike the ground surface at velocities ranging from about 10 miles per hour for small drops, up to 20 miles per hour for large drops (Laws 1941). If no plant materials cover the soil surface to intercept rainfall's energy, the impinging raindrop destroys soil aggregates by dislodging soil particles from the aggregates. These detached particles bounce around on the soil surface with the beating rains and plug large pores that, when open, promote water intake. This leads to lower water infiltration rates into the soil and greater water runoff (that is, more water flowing over the soil surface). This scenario was the case for many farm fields in June of 2008, when croplands were covered with limited residue from the previous year, and crops were not yet established to intercept rainfall's energy.

However, the second process, soil detachment due to runoff, was the major cause of soil erosion in 2008. When large, high intensity storms occurred in June (plate 4), many factors contributed to runoff and associated soil erosion and sediment transport: the soil profile was already filled with water, soil had been recently tilled, and plants, if any, were quite small. Adding heavy rains to the Iowa landscape in June was similar to pouring water into a full bucket—the bucket ran over. Rain fell on saturated soils, and much of it had no option but to run off the land (plate 5).

Runoff's capacity to transport sediment, or soil particles, depends on both flow velocity and sediment characteristics. When the sediment transport capacity of water is reduced, for example when flow velocity slows, sediment may be deposited. This usually happens when the slope of a channel is reduced, perhaps at the base of a slope, or because of an obstruction (e.g., filter strip, fenceline, terrace) in the flow path. Within agricultural fields, soil removed from one place may be deposited in another, creating a long-term productivity loss on the upslope areas from where soil was removed (as illustrated in figure 16-1) and a short-term productivity loss downslope, where crops have been buried by sediment. The sediment remaining in the flowing water that leaves the field ends up in ditches, rivers, lakes, or even flooded houses. Worth noting, sediment deposition in fields typically occurs in the same place from year to year.

In addition to within-field soil movement and deposition of sediment, floodwater-transported soil particles can cause problems when streams rise and overtop levees or simply flow out of their banks, causing sand and other coarse sediment to be deposited on flooded lands. These deposits may bury and kill plants. In subsequent years, the sand can reduce yield due to lower soil water-holding capacity and potentially lower nutrient content. The flood-water itself also can overwhelm crops and destroy them, and water quality is degraded as the chemicals, fertilizer, and manure from the field are washed into a stream.

The largest erosion events occur when the soil is bare, soil moisture is high, rainfall intensities are high, and rainfall amounts are large. All of these factors were in play on June 12, arguably the day of the largest storm of the 2008 flood period. Just four days earlier, on June 8, there had been another significant rainfall, runoff, and soil erosion event, which was followed by three days with smaller amounts of rain. Rainfall intensities on June 12 were quite high. As a result, farmland in selected areas of eastern Iowa (including both cropland and grassland) experienced runoff amounts estimated at nearly 90 percent of total rainfall, and soil erosion within given townships was estimated to approach an astounding 50 tons per acre, ten times the tolerable level for an entire year. On the average, runoff was estimated to be slightly more than 50 percent of the precipitation—an average of 2 inches of runoff from all of Linn and Johnson Counties—and many townships in these counties had an estimated runoff well above 2 inches, some in excess of 5 inches. Soil loss was estimated to average about 3.6 tons per acre across the two counties (plate 6). Estimated runoff amounts for the entire April 12 to June 12 period were very high (plate 5), in excess of 20 inches within selected townships.[1]

Although erosion is often the dominant focus on agricultural land, other types of farmland damage are equally as serious. These include damage to conservation practices (e.g., waterways, terraces, ponds) designed to control runoff, soil erosion, and sediment delivery. In addition, damage caused by the formation of gullies and channels creates management challenges for farmers and their machinery.

A runoff amount of over 2 inches is particularly significant because agricultural terraces are commonly designed to store or transport about 2 inches of runoff from the area that drains to the terrace. When larger and more intense storms occur, the landscape is overwhelmed by the sheer amount of precipitation. Agricultural terraces may overtop and fail (figure 16-2), and poorly maintained grassed waterways and other channels may be damaged (figure 16-3). Failure of terraces is particularly significant because large volumes of water may be released, causing damage to lower terraces and multiplying downhill and offsite damage. These types of failures occurred in multiple locations in eastern Iowa during the 2008 storms.

What can we do to control soil erosion under severe, rare weather events, such as the rainfall and runoff of 2008? This is a complex, multi-faceted issue with no easy answers.

Terraces, waterways, and sediment control structures are not typically

FIGURE 16-2 As water accumulates above an agricultural terrace and then overflows, the terrace becomes vulnerable to washout. This terrace washed out during the floods of 2008, threatening terraces below it with a similar fate. *Photograph by LuAnn Rolling, U.S. Natural Resources Conservation Service.*

designed to handle extreme rainfall events, mainly because structures with increased capacity are more expensive to construct, create a greater nuisance for field operations involving large equipment, and take more land out of production. How much are farmers and taxpayers willing to spend on large structures that only rarely have positive paybacks? But on the other hand, perhaps such payments are justified since farmers with smaller conservation structures that are overwhelmed by extreme rainfall may receive emergency payments at the taxpayer's expense, even as the damaged structures negatively affect everyone who lives downstream.

An alternative to conservation structures involves converting a portion of the landscape to permanent cover. Perennial vegetation protects the soil surface by reducing the force with which raindrops strike the soil surface, thus decreasing detachment rates. When rainfall is extreme, perennial vegetation will only marginally decrease runoff, but even then it will reduce erosion dramatically. Another alternative is reduced-tillage, which has had a successful adoption

FIGURE 16-3 Rapidly flowing water across the soil surface can damage areas in which this water concentrates. Even if grass is established in these areas, gullies can begin to form and the grass waterway itself can be damaged, as shown here. *Photograph by Richard Cruse.*

rate in Iowa, or no-till farming. These practices significantly reduce runoff and soil erosion for low to moderate storms. However, once the soil profile is saturated, or during extreme rainfall events, runoff will occur even on no-till sites. Then crop residue on the soil surface can become problematic as it moves with runoff water off the field and accumulates in and around culverts, ditches, and bridge approaches.

In summary, there is no silver bullet for preventing flood-related damage from extreme rainfall events to agricultural land. The spring events in 2008, particularly in eastern Iowa, demonstrated once again that Mother Nature rules. If we use the land for production agriculture, it will take heroic measures to eliminate the damages that occur to our soils and landscapes from extreme weather events. But there are measures that we can use to reduce runoff, erosion, and associated damages. We can employ more grassed waterways and agricultural terraces designed for more intense storms, and we can better maintain them. We can increase no-till practices and perennial cover, and we can restore native vegetation on particularly vulnerable areas.

And we can look to the future. We must face the fact that nearly every year we see a greater frequency of localized damage resulting from severe storms. Measures to better control damage from such events will provide benefits nearly every year somewhere in Iowa. And, with the expected increase in severe storm frequency in Iowa due to climate change (see chapter 12), improvements in design and increased implementation of conservation practices are imperative.

Note
1. The estimates in this paragraph come from techniques outlined in Cruse et al. 2006 and can be found at IEM 2009.

References Cited

Alexander, E. B. 1988. "Rates of soil formation: Implications for soil-loss tolerance." *Soil Science* 145: 37–45.

Cruse, R., D. Flanagan, J. Frankenberger, B. Gelder, D. Herzmann, D. James, W. Krajewski, M. Kraszewski, J. Laflen, J. Opsomer, and D. Todey. 2006. "Daily estimates of rainfall, water runoff, and soil erosion in Iowa." *Journal of Soil and Water Conservation* 61(4): 191–199.

Iowa Environmental Mesonet (IEM). 2009. Ames: Iowa State University Department of Agronomy. http://mesonet.agron.iastate.edu/.

Kazemi, M., L. C. Dumenil, and T. E. Fenton. 1990. "Effects of accelerated erosion on corn yields of loess-derived and till-derived soils in Iowa." Final report for Soil Conservation Service, Agreement No. 68-6114-0-8, Des Moines, Iowa.

Laws, J. D. 1941. "Measurement of the fall velocity of water drops and raindrops." *Transactions of the American Geophysical Union* 22: 709–721.

Meyer, L. L., and J. V. Mannering. 1967. "Tillage and land modification for water erosion control." *Proceedings, American Society of Agricultural Engineering, Tillage for Greater Crop Production Conference.* December 11–12, 1967, pp. 58–62.

Montgomery, D. R. 2007. "Soil erosion and agricultural sustainability." *Proceedings of the National Academy of Science* 104: 13268–13272.

Mutel, C. F. 2008. *The Emerald Horizon: The History of Nature in Iowa.* Iowa City: University of Iowa Press.

Owens, L. B., and J. P. Watson. 1979. "Rates of weathering and soil formation on granite in Rhodesia." *Soil Science Society of America Journal* 43: 160–166.

Dana Kolpin
Keri Hornbuckle

17 What's in Your Floodwaters?

WHEN PEOPLE THINK about flooding, they recall the destructive nature of the water or, for those unlucky enough to have had homes or businesses inundated by floodwater, the muddy mess that it left behind. In many Iowa communities hit by the June 2008 floods, buildings were coated with muck and sand after the floodwater receded. That material, which we will call by the general name sediment, was everywhere the floodwaters reached. It smelled of sewage and decay and attracted flies in some neighborhoods. It collected in basements and in carpets of flooded homes and businesses. It coated the sides of the buildings, leaving a clear indicator of how high the water had risen.

However, there is another aspect of flooding that receives little public attention—namely, the pollutants that are present in the water and sediment during such extreme flow conditions. It is easy to understand why pollutants in these floodwaters are often ignored. Such pollutants are invisible to the eye and their potential effects are far subtler than the obvious effects of the water and sediment. In this chapter, we discuss the toxic or potentially toxic chemicals that the flood may have transported. While few data are available yet about exactly what chemicals were in the historic 2008 floodwaters, previous research on the Cedar River and other Iowa streams provides helpful clues to the probable floodwater contaminants (Goolsby et al. 1993; Kolpin et al. 2000; Schnoebelen et al. 2003).

So where do rivers pick up pollutants during times of flooding? Because Iowa is known as the tall corn state, it's easy to comprehend that farm fields are a major source of a variety of floodwater pollutants (figure 17-1). Whatever is applied to fields or stored on farms is available to be washed off into rivers and streams. Possible pollutants include sediment, nutrients from animal manure and commercial fertilizers, and a wide range of pesticides that are used to control weeds, insects, and fungi. Given that Iowa has a much larger population of farm animals than people, it also should be of no surprise that livestock production is another potential source of pollutants during flooding. Possible pollutants washed from animal feeding operations into floodwaters include bacteria, nutrients, and hormones that are naturally associated with animal manure, and veterinary drugs such as antibiotics and synthetic hormones

FIGURE 17-1 Heavy rains on farm fields during the June 2008 flooding washed topsoil and recently applied pesticides into local streams and rivers. *Photograph by Dana Kolpin, taken June 14, 2008.*

that may be present in manure because of their use to promote growth and reduce disease.

A wide range of pesticides were applied to agricultural soils in the weeks before the 2008 flooding occurred. Because the heavy winter snows led to a wetter than normal spring, many farmers were not able to get into the fields until late spring. As a result, many acres of cropland had been treated with pesticides just before the rains and floods occurred in early June. In Iowa, the pesticides that may have been washed into rivers and creeks include acetochlor, atrazine, glyphosate, and metolachlor (Owen and Hartzler 2008). The loss of these chemicals from their intended soils was expensive for farmers, as many needed both to replant the fields and reapply chemicals. In addition, the presence of such chemicals in streams can be harmful to aquatic plants and animals downstream.

As well as the agricultural sources of pollutants during flooding, there is a complex mixture of potential urban sources in both residential and industrial areas. Because entire towns were inundated during the record 2008 flooding, the contents of homes and industrial buildings were washed into rivers and creeks. Considering all the items stored in the average garage (such as paints, cleaning supplies, and lawn and garden chemicals), it's not hard to imagine the myriad of pollutants that could be derived from urban sources.

Another potential urban pollutant source is wastewater treatment plants, which are commonly located close to rivers and may be overwhelmed by storm-water flow or directly damaged by floodwater (figure 17-2). In Cedar Rapids, for example, the wastewater treatment plant was severely damaged by flooding of the nearby Cedar River and was not able to provide basic treatment for more than five weeks after the flood, during which time untreated sewage was discharged into the river. Advanced treatment was not functional even three months after the floodwater receded. Even treated municipal waste, however, is known to contain a host of urban-derived pollutants (Glassmeyer et al. 2005) including pharmaceuticals (such as antibiotics, pain relievers, antidepressants, and birth control hormones), personal care products (such as fragrances, detergents, and components of antimicrobial soaps), and other products that we use in our everyday lives (such as caffeine and insect repellents). These pollutants have been found in our nation's surface waters and in Iowa's waters (Kolpin et al. 2002). Even under normal performance, typical wastewater treatment plants are not able to completely remove these compounds. Because of the number of wastewater treatment plants affected by the June 2008 floods, it is anticipated that large amounts of these chemicals were released into many Iowa rivers

FIGURE 17-2 Excessive stormwater flow and flooding of municipal wastewater plants made treatment of sewage difficult during the June 2008 flood and resulted in raw or minimally treated wastewater being discharged into local streams and rivers. This June 14 photograph shows the Iowa River submerging a section of Highway 6 in Iowa City, including the north Iowa City wastewater treatment plant. *Photograph by Dave Schwarz,* Iowa City Press-Citizen.

and streams. The effect of such a chemical cocktail in our rivers and streams is unclear but certainly worrisome for both humans and wildlife that come in contact with these waters.

Cedar Rapids is home to a variety of industries, including metal foundries, metal-plating facilities, meat packing plants, grain mills, paper and packaging plants, food processing plants, and coal-fired electric power generation plants. Many of these facilities were located within the reaches of the June 2008 flood. Figure 17-3 shows the location of these industrial facilities within the flooded region of Cedar Rapids. Many of the older industries in Cedar Rapids may have stored potentially toxic chemicals in the past or may have been constructed using hazardous materials, including polychlorinated biphenyls (PCBs), a particularly persistent pollutant that has been banned since 1979 but was historically used in a wide variety of industrial applications (ATSDR 2000). In addition, the destructiveness of the June 2008 floods may have introduced PCBs back into the environment when transformers cracked or collapsed or when buildings were demolished. At high levels of exposure PCBs may cause cancer, neurotoxicity, and disturbances in the hormonal systems of animals and humans.

Although determination of the chemicals in water and sediment samples collected during the historic 2008 flooding is still under way, there are some

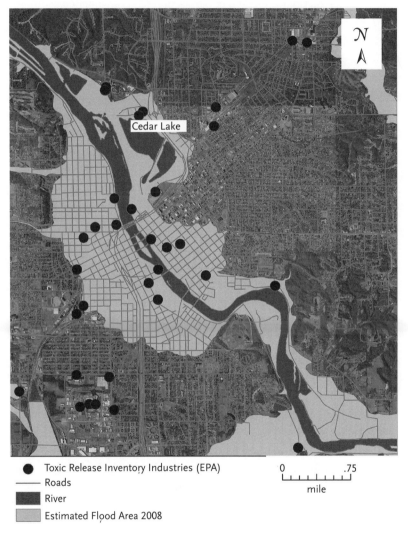

FIGURE 17-3 The extent of the June 2008 flood of the Cedar River at Cedar Rapids, with the Cedar River's maximum extent during the flood shown in light gray. The plotted circles indicate the location of industrial facilities in the city. The location of Cedar Rapids is shown in figure I-2. *Illustration by Andres Martinez, University of Iowa, with base map from Linn County Auditors. Sites from EPA's Toxic Release Inventory.*

early data available to help put flood-related water-quality questions into context. On the afternoon of June 14, 2008, personnel from the U.S. Geological Survey in Iowa City measured the flow of the Cedar River at the Highway 30 bridge just downstream from Cedar Rapids and collected a series of water samples to be analyzed for a variety of pollutants. At this time, the Cedar River was flowing at a rate of 116,000 cubic feet per second (cfs). Suspended sediment had a measured concentration of 175 milligrams per liter (parts per million or ppm), nitrate was at 5.3 ppm, and atrazine had a concentration of 2.27 micrograms per liter (parts per billion). The flow of the Cedar River at this time was about 53 times the normal flow of the Cedar River (2,200 cfs) over the 105 years of record at this site (USGS 2008). To appreciate the magnitude of such concentration levels, the quantity of pollutants being transported by the Cedar River can be calculated by multiplying these chemical concentrations by the flow in the Cedar River. Thus, at the time these water samples were collected, 1,268 pounds of suspended sediment, 38 pounds of nitrate, and 0.016 pounds of atrazine flowed downstream past the Highway 30 bridge each second. Put another way, in a single minute, the Cedar River was transporting enough pollutants to fill two and a half 15-ton dump trucks with sediment; fertilize 20 acres of cropland with nitrate; and treat 1 acre of corn with atrazine. These data make it clear that over the course of the week-long flood event, very large amounts of sediment, fertilizer, and human-derived chemicals passed through Cedar Rapids, and the amounts of pollutants being transported by the 2008 flood were tremendous indeed.

The impacts of the floodwaters' chemical content and transport are not clear, but consider the intended purpose of many of the pollutants. For example, pharmaceuticals were designed to have a physiological effect in animals and/or humans, and pesticides were designed to affect insects and plants in farm fields, lawns, and gardens. Industrial chemicals like PCBs, on the other hand, were not designed to have any biological effect at the time they were produced but much later were discovered to have harmful toxic properties. None of these compounds were intended to be released into the environment in the manner and magnitude that occurred during the 2008 floods. Thus, there are many questions about the overall impact of floods on our natural environment. The effect of human-made chemicals being transported and deposited during flooding is certainly among those questions deserving further attention.

References Cited

Agency for Toxic Substances and Disease Registry (ATSDR). 2000. *Toxicological Profile for Polychlorinated Biphenyls (PCBs)*. Atlanta, Ga.: U.S. Department of Health and Human Services.

Glassmeyer, S. T., E. T. Furlong, D. W. Kolpin, et al. 2005. "Transport of chemical and microbial compounds from known wastewater discharges: Potential for use as indicators of human fecal contamination." *Environmental Science & Technology* 39: 5157–5169.

Goolsby, D. A., W. A. Battaglin, and E. M. Thurman. 1993. *Occurrence and Transport of Agricultural Chemicals in the Mississippi River Basin July through August 1993*. U.S. Geological Survey Circular 1120-C. http://pubs.usgs.gov/circ/1993/circ1120-c.

Kolpin, D. W., E. E. Fischer, and D. J. Schnoebelen. 2000. *Water-quantity and Water-quality Aspects of a 500-year Flood—Nishnabotna River, Southwest Iowa, June 1998*. U.S. Geological Survey Water-Resources Investigations Report 00-4025. http://ia.water.usgs.gov/pubs/reports/WRIR_00-4025.pdf.

Kolpin, D. W., E. T. Furlong, M. T. Meyer, et al. 2002. "Pharmaceuticals, hormones, and other organic wastewater contaminants in U.S. streams, 1999–2000: A national reconnaissance." *Environmental Science & Technology* 36:1202–1211.

Owen, M. D. K., and R. G. Hartzler. 2008. *2008 Herbicide Guide for Iowa Corn and Soybean Production*. Ames, Iowa: Cooperative Extension Service, Iowa State University.

Schnoebelen, D. J., J. J. Kalkhoff, K. D. Becher, and E. M. Thurman. 2003. *Water Quality Assessment of the Eastern Iowa Basins: Selected Pesticides and Pesticide Degradates in Streams, 1996–98*. U.S. Geological Survey Water–Resources Investigations Report 03-4075. http://pubs.usgs.gov/wri/2003/wri034075/.

U.S. Geological Survey (USGS). 2008. U.S. Geological Survey, National Water Information System. http://waterdata.usgs.gov/nwis.

Peter S. Thorne

18 Air Quality Hazards

DURING TIMES OF flooding, waterborne hazards are obvious, but few think about the dangers that might be drifting through the air. Yet floods produce airborne hazards that often linger long after the flooding has ended.

Floods produce airborne chemical health hazards when previously contained toxic substances are released into the environment, and microbial hazards when post-flood conditions promote the growth of microorganisms. Some airborne hazards, like carbon monoxide, are widely recognized. Others such as spores from the mold *Aspergillus niger* may be unfamiliar to the general public.

Exposure to airborne hazards can occur in a variety of manners. Toxic agents carried and deposited by floodwaters can later become airborne through volatilization (evaporation) or through attachment to airborne dust particles. Hazardous agents also can appear when floodwaters recede. This is the case for molds that grow primarily after flooding. Still other toxicants, like asbestos and lead, cause problems only after a flood, when buildings are demolished.

Dangerous chemical exposures can occur during flood recovery activities such as cleaning and demolishing buildings. Exposures arise from sludge and debris or from chemicals used in the cleaning process such as bleach, detergents, and biocides. Previously contained toxic substances may be released during building demolition. One such toxicant is asbestos, which may be released from insulation, pipe wrapping, and floor tiles. In the aftermath of a flood, containment practices normally employed during asbestos abatement may be ignored. Exposure to lead dust from lead-based paint also may pose an increased risk for workers or people living in the area of the demolition. My research team performed investigations in Cedar Rapids to measure exposures to lead and asbestos during post-flood building demolition or renovation. We collected air samples onto fine filters using a vacuum pump and then analyzed the filters by various scientific approaches, including microscopy techniques for asbestos and very sensitive analytical chemistry processes for lead and other trace metals. This work will help us understand human health risks associated with flood recovery.

Other chemical hazards may be created by the post-flood need for power. Because electrical systems are often impaired or completely non-operational, the use of gasoline-powered portable generators and tools such as power washers, blowers, heaters, and air compressors becomes widespread. Even though most people know better, these generators and gasoline-powered tools often are used indoors, which can lead to carbon monoxide (CO) poisoning. There are 500 deaths in the U.S. each year from CO poisoning, and flood-related exposures contribute to this toll (CDC 2005a; CDC 2006). Carbon monoxide can be monitored with devices that provide a continuous readout or sound an alarm when an unsafe level is reached. However, most often the appearance of symptoms is the first evidence of CO poisoning. On June 25, 2008, thirteen workers were using gas-powered equipment without proper ventilation to clean a flood-damaged building in downtown Cedar Rapids. They experienced lightheadedness, dizziness, and nausea. Firefighters measured high levels of CO at the scene. The workers were treated at a hospital and all recovered. Just one week earlier, 16 Mercy Medical Center employees were evaluated for CO poisoning when exhaust fumes entered through an air intake. Diesel generators powering large dehumidifiers had been placed too close to the buildings. The generators were moved farther from the building, and that air intake was temporarily closed. The workers all recovered.

In most floods, chemicals such as gasoline, pesticides, and industrial chemicals that have settled onto the soil may be volatilized when the land begins to dry out. This was a problem in some low-lying areas after Hurricane Katrina (Manuel 2006). In New Orleans, this problem was increased by the large petrochemical industrial sector, the presence of a large number of Superfund toxic waste sites, and the flat topography that held rather than shed the floodwaters and their pollutants. However, in the 2008 Iowa flood, huge volumes of flowing water provided sufficient dilution of most of these chemical hazards so that human health problems probably were limited.

In the wake of disasters, microorganisms often produce infectious diseases (CDC 2005b). Infectious hazards may arise when people are suddenly clustered in large groups in shelters, sometimes under poor hygienic conditions. Recall the horrific conditions in the Louisiana Superdome, the "shelter of last resort" for stranded New Orleans residents after Hurricane Katrina. Here 20,000 people spent six or seven days in sweltering heat with no power, limited food and water, no cots, no functioning toilet facilities, and no designated medical staff (Thevenot and Russell 2005). Such conditions can promote person-to-person transmission of infectious viruses and bacteria, for example those causing influenza, acute viral gastroenteritis, and tuberculosis. Thus, it is advantageous to place people in smaller shelters and to minimize the length of their stays. At its peak, the flood in Cedar Rapids displaced around 10,000 residents who found temporary shelter in schools, hotels, hospitals, and with relatives and friends (Hart 2008). Fortunately, there were no significant airborne infectious-disease outbreaks because of low pre-flood infectious-disease levels, the use of multiple smaller shelters, and adequate hygiene and basic services.

Airborne microorganisms also cause non-infectious illnesses, most commonly exacerbation of mold allergies and asthma, and organic dust toxics syndrome (ODTS, also known as toxic pneumonitis) (Douwes et al. 2008). Molds and bacterial endotoxins, ubiquitous during floods, are the principal causes of these conditions. Endotoxins are molecules that make up bacterial cell membranes and are potent inflammatory agents in the lung. Bacteria and endotoxins are found in floodwaters and are left behind when flooding subsides. One month after Hurricane Katrina struck the Louisiana and Mississippi coast in 2005, a group of collaborating researchers and I collected and analyzed air samples from flooded homes in which people were living and working. In these homes, we measured endotoxin concentrations in the air that were as high as concentrations in livestock barns, where extremely high levels are typical (Chew et al. 2006). Inhalation of endotoxins produces symptoms that include fever, achiness, fatigue, and airway inflammation, leading to wheezing and asthma (Thorne and Duchaine 2007). Molds are fungi that grow anywhere they find a substrate for growth, adequate dampness, a suitable temperature range, and the absence of direct sunlight. Flooded buildings satisfy all these needs. These homes contained extremely high concentrations of molds, demonstrating that they were not safe for occupancy (Chew et al. 2006).

Molds thrive in flooded environments. They are saprophytic organisms, acquiring the nutrients they need from decaying organic matter. The majority of molds prefer temperatures between 60 and 86°F and sufficiently damp surfaces in which the air is at least 80 percent saturated. When airborne mold spores land on a damp surface, they begin to grow mycelia (vegetative thread-like fibers) that penetrate the surface's exterior and draw nutrients to support their growth (figure 18-1). If the material is a book, the molds break down cellulose in the paper for nutrition. Molds growing on wallboard digest the cellulose outer layer and starches contained in the gypsum core. Molds growing on concrete block walls use surface dirt and grime for nutrition. As molds grow, they produce volatile organic compounds, some of which are foul-smelling. When the mycelia have spread through the organic matter, they sprout structures called conidiophores

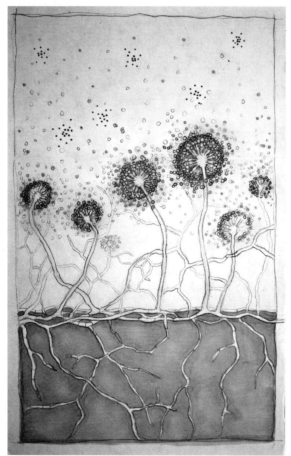

0 ———— 50
micrometers

FIGURE 18-1
Aspergillus mold growing in damp material such as wet wallboard. Mycelia grow within the material and give off odorous microbial volatile organic compounds. Sporulating structures (conidiophores) are shown releasing spores that further disseminate the mold and can be inhaled, causing illness. *Illustration by Jeanne DeWall.*

that produce spores: small reproductive structures that disperse easily. Spores can survive for extended periods under harsh conditions. When released into air currents, spores further spread the molds. Some types of mold produce few spores that are large in size and do not remain airborne. However, many of the molds that arise in wet buildings, such as *Aspergillus* and *Penicillium* species, produce thousands of spores from each conidiophore. Each spore is about 10 microns in diameter (10 millionths of a meter), and 36,000 would be required to cover the head of a pin. That makes them perfectly sized for penetrating the nose, trachea, and lung airways.

After Iowa floodwaters receded, mold growth accelerated, driven by the damp conditions and warm Iowa summer weather. Carpets, books, clothes, furnishings, wallboard, wood, and even concrete walls became covered with mold. Figure 18-2 shows a leather chair and footstool in a home that was flooded with 2.5 feet of standing water. This chair and other furnishings were completely covered with a variety of molds including *Aspergillus*, *Penicillium*, *Trichoderma*, *Zygomycetes*, and *Paecilomyces*. Several types of mold were growing on the wet wallboard behind the chair, especially *Aspergillus* species. These molds are allergens for many people.

When mold spores and mycelia fragments are inhaled, they interact with lung immune cells and cause inflammation and, in some people, allergic responses. Symptoms include nose and throat irritation, cough, malaise, fatigue, and headache. Itchy red eyes, skin rashes, and recurrent sinus infections are also common. Relief workers in areas flooded by Hurricane Katrina referred to this condition as the Katrina cough or Katrina flu. Some exposed people develop mold allergies, and some develop potentially life-threatening allergic asthma. Individuals who are already allergic to molds or other common allergens are particularly susceptible. Mold exposures may trigger attacks of asthma with shortness of breath, wheezing, and cough that can interfere with activities of daily life and significantly reduce the quality of life. This condition may necessitate lifelong health monitoring and daily administration of several different medications.

Very high levels of exposure to bioaerosols are possible when removing debris from flooded buildings or performing interior demolition. Under these circumstances, exposed persons may develop ODTS or hypersensitivity pneumonitis. ODTS is an acute illness characterized by fever, dry cough, chest tightness, shortness of breath, headache, fatigue, and general malaise. It appears within hours of exposure and is usually limited to two days, as long as exposure is discontinued. This disease does not require prior allergic sensitization and is

FIGURE 18-2 A leather chair and footstool are shown in a flooded home after the water has receded. The line on the wall marks the height reached by the water. A variety of different molds have engulfed the chair, end table, wall, and books on the shelf. *Photograph by Christine A. Rogers.*

most often caused by exposure to a combination of endotoxin and fungal spores. Hypersensitivity pneumonitis is a less common allergic disease. Symptoms are nearly the same as those of ODTS, but the disease can persist and develop into a chronic form that can lead to impaired breathing.

As molds proliferate, they produce volatile compounds that have a very strong and unpleasant odor. These odors, variously described as putrid, fetid, or like rotting vegetables, can make exposed people feel nauseous or gag.

During and after the Iowa flood, many but not all people properly protected themselves while working in mold-contaminated areas. A sample of the recommendations I provided to the public is listed in table 18-1. These guidelines and instructions on how to clean mold from homes and possessions were disseminated by volunteer groups and through television, print media, and websites. We also produced a six-minute video entitled *Mold Hazards in Flooded Buildings* and posted it on various websites and YouTube. These educational

TABLE 18-1 Guidance Disseminated to the Public to Warn about Mold Hazards after the Flood

---

How can you protect yourself while working in a mold contaminated area?

- Children, pregnant women, the elderly, people with asthma, and anyone with an immune system disease should not be in the area when repairs and cleaning are taking place.
- Wear protective clothing, waterproof gloves, protective eyewear, and a disposable shower cap (to keep mold spores out of your hair).
- Use a tight-fitting, two-strap N95 mask (respirator) when cleaning up mold. A STANDARD DUST MASK OR SURGICAL MASK WILL NOT PROTECT YOU. Ensuring a proper fit and replacement of the mask are necessary to reduce mold exposure.
- Do not attempt to clean large areas of mold. If you start cleaning and your health starts changing, get out! Call a professional.
- Be very cautious when cleaning mold because mold releases spores when it is disturbed. The area should be well ventilated when you clean, but do not use a fan. Fans may spread existing mold and send more spores into the air.
- Avoid touching your face, especially mouth and eyes. Hands should be washed thoroughly before eating.
- If there is a large amount of mold a mask will only help a little and you will still be exposed.

---

measures hopefully reduced Iowa's incidence of Katrina cough and other mold-induced diseases.

Even several months after the flooding, as I walked through Cedar Rapids neighborhoods affected by the flood, it was apparent that occupied houses were interspersed among condemned and abandoned homes. A family could be living in a home that was in disrepair, while an adjacent abandoned home had not been touched since the flood (figure 18-3). Such homes were often full of mold, and some had floodwater in the basement for an extended period. Toxicants from adjacent abandoned homes can pose significant hazards, and nearby residents may face exposures to asbestos, lead dust, endotoxin, and mold spores. My research team performed environmental monitoring during demolition and building renovation to help us determine the extent of these hazards. This work demonstrated that significant exposure hazards exist but can be mitigated with proper risk evaluation and public health intervention.

In conclusion, even after flooding has ended and river levels have returned to normal, there are dangerous airborne exposures to chemical and biological hazards. Increased awareness of the hazards and effective public health measures can reduce the risk of harm.

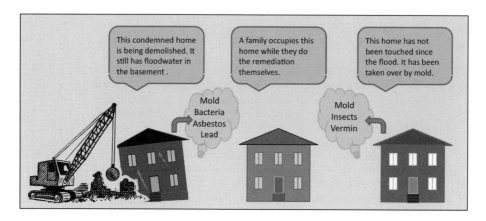

FIGURE 18-3 For many months after the flooding of Cedar Rapids, families lived in flood-damaged homes adjacent to condemned homes and homes that had not been touched since the flood. This posed the risk of dangerous exposures to asbestos, lead, molds, and possibly other hazards. *Illustration by the author.*

References Cited

Centers for Disease Control and Prevention (CDC). 2005a. "Carbon monoxide poisoning after Hurricane Katrina—Alabama, Louisiana, and Mississippi, August–September 2005." *Morbidity and Mortality Weekly Report* 54: 996–998.

————. 2005b. "Infectious disease and dermatologic conditions in evacuees and rescue workers after Hurricane Katrina—multiple states, August–September 2005." *Morbidity and Mortality Weekly Report* 54: 1–4.

————. 2006. "Carbon monoxide poisonings after two major hurricanes—Alabama and Texas, August–October 2005." *Morbidity and Mortality Weekly Report* 55: 236–239.

Chew, G. L., J. Wilson, F. A. Rabito, et al. 2006. "Mold and endotoxin levels in the aftermath of Hurricane Katrina: A pilot project of homes in New Orleans undergoing renovation." *Environmental Health Perspectives* 114: 1883–1889.

Douwes, J., W. Eduard, and P. S. Thorne. 2008. "Bioaerosols." In *International Encyclopedia of Public Health, Vol. 1*, ed. K. Heggenhougen and S. Quah, pp. 287–297. San Diego: Academic Press.

Hart, Tom. 2008. Environmental Specialist, Linn County Health Department, Cedar Rapids, Iowa. Personal communication, September 22, 2008.

Manuel, J. 2006. "In Katrina's wake." *Environmental Health Perspectives* 114: A32–A39.

Thevenot, B., and G. Russell. 2005. "Reports of anarchy at Superdome overstated." *Seattle Times*, September 26, 2005.

Thorne, P. S., and C. Duchaine. 2007. "Airborne bacteria and endotoxin." In *Manual of Environmental Microbiology, 3rd Edition*, ed. C. J. Hurst, R. L. Crawford, J. L. Garland, D. A. Lipson, A. L. Mills, and L. D. Stezenbach, pp. 989–1004. Washington, D.C.: ASM Press.

John Pearson

19 Flood Effects on Natural Communities

FLOODS BRING TREMENDOUS damage to human communities in the form of swamped homes, ruined businesses, inundated farms, wrecked bridges, and devastated lives, so it is difficult to imagine that the same floods might renew and restore natural communities of plants and animals. But while the water that fills a basement is destructive, water that fills a dried wetland creates habitat for ducks, frogs, turtles, salamanders, dragonflies, and waterlilies. Sand deposited across farm fields, parks, and roads must be painstakingly removed to restore their usefulness, but fresh sand deposits in undeveloped bottomlands provide new opportunities for pioneering plants—rare in mature forests and prairies— to become established and to thrive (figure 19-1). Bridges are undermined by streambed erosion and are further stressed by logjams accumulating on their upstream sides, but the scouring of holes in the river bottom and lodging of uprooted trees (snags) along streambanks build a diversity of habitats that sustain fish and mussels. Unlike human communities, natural bottomland communities are adapted to periodic floods and would eventually disappear without them.

The flood of 2008 fulfilled its dual role as an agent of destruction of human-

FIGURE 19-1 Sand deposit on a floodplain of the Cedar River near Palo, September 2008. Pioneering plant species colonized this new habitat shortly after the flood. Similar deposits in farm fields and urban areas must be laboriously removed. *Photograph by the author.*

built infrastructure and agent of renewal in bottomland habitats and riverine natural areas. During a September field trip to conservation areas near Palo (a small Linn County town hard hit by Cedar River flooding), I saw abundant evidence of natural resilience and renewal: large silver maples standing unscathed despite water marks far up their trunks, swamp white oak saplings sprouting new leaves to replace drowned foliage, asters blooming amongst grasses and sedges on recent sand deposits, flood-emplaced snags forming fish shelters along formerly featureless banks, spacious sandbars attractive to foraging shorebirds and nesting turtles (figure 19-2), oxbow lakes brimming with water, and glimpses of freshly scoured holes in the riverbed holding mussels washed from upstream.

But in addition to its expected benefits, did the flood of 2008 bring unexpected harm to the natural environment? It was unusually big and carried a distinctly unnatural load of urban debris, sediment from cropland, and chemicals from

FIGURE 19-2 Newly formed sandbar on the Cedar River near Bertram, November 2008. Large, open sandbars like this one provide habitat for shorebirds and nesting turtles. *Photograph by the author.*

farms and cities. Moreover, it flowed through a fragmented landscape whose natural habitats have been reduced to narrow corridors and small, scattered patches. Was the flood so big, so dirty, and so disruptive of remnant natural areas that it compromised the historically restorative role of flooding?

The 2008 floods on the Cedar and Iowa Rivers were indeed large, in fact the largest on record in Linn County and among the largest in Johnson County. This means that the crest of this year's flood reached land that is not often flooded, affecting species and communities that are not as flood-adapted as their streamside counterparts. Examples include forest trees on elevated terraces, prairie remnants high on the floodplain, and terrestrial organisms such as the ornate box turtle. Walnut, hackberry, and basswood often grow on terraces above the lowest floodplains and are less tolerant of flooding than silver maples, cottonwoods, box elders, and willows, which thrive in the frequently flooded portions of bottomlands. Following the 1993 floods, many walnuts died because of saturated soil and the burial of their roots under sediment (Vitosh 2008). Mortality of affected trees is not always immediate, so loss of walnuts, hackberries, and basswoods may not be evident until future years. Regeneration and growth of new seedlings and saplings can replace old trees that have died, but when growing in dense forests, young trees must be tolerant of the semishaded condition of gaps left by dead trees. Basswood and hackberry are shade-tolerant and would be expected to regenerate without difficulty, but walnuts are intolerant of shade and may fail to reestablish in mature forests.

Remnants of the native prairie that originally covered Iowa are rare, but one of the most well known is Williams Prairie in Johnson County, dedicated as a state nature preserve since 1976 (Herzberg and Pearson 2001). Located at the southern edge of the broad, flat valley of the Iowa River and containing nearly 300 native plant species, most of the preserve consists of a sedge meadow in a gently sloping, low-lying swale. Since 1956, the swale has been included in a flowage easement intended to accommodate inundation from the most extreme flood pool of the Coralville Reservoir. It has experienced ponding from high reservoir levels in two years: 1993 and 2008. In a normal year, the soil is saturated and may have shallow standing water in the spring and early summer. However, in 2008, with the reservoir fully occupying its designated flood pool, the swale held water up to 5 feet deep that pooled over parts of the sedge meadow for up to two weeks (Corps 2008). Although sedge meadow vegetation is tolerant of natural wetland conditions, this rare plant community is not adapted to deep inundation. The meadow also was flooded in 1993, but unfortunately flood impacts were not studied then. Observations will be needed in future years to determine what effects the inundation of 2008 (now confounded with the 1993 event) may have had.

Ornate box turtles (figure 19-3) are adapted to a fully terrestrial lifestyle (Christiansen and Bailey 1988). Classified as a threatened species in Iowa, these box turtles are almost entirely restricted to small, scattered patches of sandy upland soil, mostly in the eastern half of the state. Although they can enter water, their high-domed shell contrasts sharply with the flattened, streamlined form of aquatic turtles, making them inefficient swimmers. Living in uplands normally free from flooding, box turtles were nonetheless subjected to the 2008 flood when the rising Iowa River invaded their habitat in the Hawkeye Wildlife Area upstream from the Coralville Reservoir. There, turtles were presumably herded onto ever-shrinking hilltop islands as floodwaters threatened to inundate the rolling sandy landscape. Fortunately, the flood receded before the hilltops were completely submerged, presumably allowing the turtles to

FIGURE 19-3 Ornate box turtle, *Terrapene ornata*, a rare species found in sandy soils in eastern Iowa. *Photograph by Neil Bernstein.*

survive. However, despite searches in June and July (including a trip by kayak to the isolated hilltops during peak high water), researchers did not observe any turtles in the summer of 2008 (Bernstein 2008). But hot temperatures normally drive the turtles into burrows by midsummer, so their lack of detection at that time could not distinguish between normal circumstances and flood-related effects. As with terrace trees and sedge meadow vegetation, assessing the survival of box turtles will require additional study.

Aquatic life in streams, including fish, mussels, and insects, is most diverse and abundant where a variety of substrates—cobbles, gravel, sand, silt, and clay—is present. But the immense loads of sediment transported by floods tends to be deposited as long homogeneous strands of single classes of sediment. This reduces habitat diversity and limits the diversity of aquatic organisms. Sand, in particular, often dominates the beds of the Cedar River and the Iowa River below the Coralville Reservoir, forming a shifting substrate poor for the spawning of fish and survival of mussels. Uniform habitats favor common, generalist species of fish such as carp and mussels such as plain pocketbooks, while working against the success of rarer, more specialized species of fish such as daces and darters, and mussels such as pistolgrips and hickorynuts (Gritters 2008; Poole 2008). Heavy flood-related precipitation eroded much soil from cropland and delivered a great influx of sediment to the Cedar and Iowa Rivers. Although uniformity of substrate is a chronic problem in these rivers, the flood of 2008 worsened the situation by depositing still more sediment and covering rocky riffles in the Iowa City area with sand (Sleeper 2008).

Floodwaters invaded farms, business districts, neighborhoods, and industrial parks, acquiring a freakish mélange of anything that could be floated or dissolved. In addition to coarse physical debris such as tires, refrigerators, steel drums, plastic containers, and trash posing entanglement hazards to wildlife, this concoction included contaminants capable of threatening water quality and aquatic life, including sewage, fertilizer, oil, pesticides, and industrial chemicals. However, the staggering volume of floodwater greatly diluted these pollutants. Other than bacteria (which rose during the flood due to runoff, and remained high for much of the summer downstream of cities with disabled wastewater treatment plants), concentrations of contaminants—including nitrates, oil, herbicides, volatile compounds, and metals—were low (Skopec 2008). But in contrast to the main river channel, the quality of water and sediments trapped in non-flowing backwaters may not have been as benign. Unfortunately, backwater areas—including off-channel wetlands and oxbow lakes vital to many species of wildlife—have not been sampled, so their status remains uncertain. Finally, although concentrations of contaminants were low in Iowa rivers, total loads were high (see chapter 17), effectively transporting the problem to downstream water bodies and their inhabitants and ultimately to the Gulf of Mexico.

Effects of flooding on natural communities may be magnified by the reduced area and fragmented distribution of their habitats on the modern landscape. In 1840, when surveyors with the General Land Office drew the first maps of the Cedar and Iowa valleys in Linn and Johnson County, tallgrass prairie prevailed on the broad, gently rolling uplands between the major rivers, comprising nearly 70 percent of the landscape (plate 7). Forest made up most of the remaining 30 percent, occupying bottomlands, bluffs, and slopes. Land cover today is dominated by human uses. Cropland in the two-county area takes up nearly half (49 percent) of the landscape, while urban areas and roads occupy an additional 7 percent (plate 8). Grassland (now mostly pastures and hayfields dominated by non-native grasses instead of original tallgrass prairie) and forest (now largely comprised of disturbed, second-growth stands) make up less than half of their original extent.

Modern land use also has fragmented natural areas into small, scattered patches and narrow strips separated by expanses of cropland and urban environments. Fragmentation affects remnant natural areas even in normal years, but large floods exacerbate the problem. A map of the maximum flood extent shows that many natural areas, especially forest, were inundated (plate 9). In the pre-settlement landscape, wildlife displaced from bottomlands would simply move to uplands until waters receded. In today's landscape, animals moving to higher ground may find themselves in conflict with humans, houses, and highways.

Another problem posed by habitat fragmentation is the inability of sedentary species to recolonize sites that have been vacated. In the pre-settlement landscape, distances between similar habitats were short, allowing even small animals to move into vacated territory, but the elimination of "stepping-stone" patches now prevents dispersal across large expanses of inhospitable land use. If ornate box turtles, for example, are eliminated from the Hawkeye Wildlife Area, the species could permanently disappear from this site because other box turtles are too far away to immigrate from distant sand prairies. This concern could also be extended to rare native plant species in tiny natural remnants like Williams Prairie, which is far removed from other native sedge meadows.

In closing, keeping with their long history with floods, natural bottomland communities demonstrated resilience and renewal to flooding in 2008, but there were also elements of uncertainty associated with this event. There may be fewer walnuts on terraces in coming years, and sand now dominates riverbed habitats more strongly than before. Due to dilution from floodwaters, effects on Iowa streams from contaminants appear to be minimal, but large contaminant loads were exported downstream. The jury is still out on the questions of backwater contaminants, floristic losses at Williams Prairie, and the fate of ornate box turtles in the Hawkeye Wildlife Area. As an ecologist, I hope for a future that allows natural bottomlands to continue to receive the benefits of periodic flooding. The 2008 flood demonstrates the need to manage our fragmented and increasingly human-dominated landscape in ways that preserve natural habitats not just under normal circumstances but in extraordinary ones as well.

Acknowledgments

I thank Neil Bernstein, Dennis Goematt, Scott Gritters, Daryl Howell, Bernie Hoyer, John Olson, Kelley Poole, Paul Sleeper, and Mark Vitosh for providing information about fish, mussels, and turtles, guiding me to interesting sites in the field, or reviewing earlier drafts of my manuscript. I also thank Casey Kohrt for preparing the landcover maps.

References Cited

Anderson, P. F. 1996. "GIS research to digitize maps of Iowa 1832–1859 vegetation from General Land Office township plat maps." Report to State Preserves Advisory Board, Iowa Department of Natural Resources, Des Moines, Iowa.

Bernstein, Neil. 2008. Professor of Biology, Mount Mercy College, Cedar Rapids. Personal communications, September–November 2008.

Christiansen, J .L., and R. M. Bailey. 1988. *The Lizards and Turtles of Iowa*. Nongame Technical Series #3. Des Moines: Iowa Department of Natural Resources.

Gritters, Scott. 2008. District Fisheries Biologist, Iowa Department of Natural Resources, Guttenberg, Iowa. Personal communication, November 2008.

Herzberg, R., and J. Pearson. 2001. *The Guide to Iowa's State Preserves*. Iowa City: University of Iowa Press.

Kollasch, P. 2008. Digital data sources for 2002 land cover and 2008 flood extent. Natural Resources Geographic Information Systems Library, Iowa Geological Survey, Iowa City, Iowa. http://www.igsb.uiowa.edu/nrgislibx/.

Poole, Kelly. 2008. Malacologist, Limnology Laboratory, Iowa State University, Ames, Iowa. Personal communication, November 2008.

Skopec, M. 2008. "Floodwater and sediment monitoring." In *Rebuild Iowa Office Agriculture and Environment Task Force, Supplemental Information to August 2008 Report*. www.rio.iowa.gov/task_forces/ag-enviro/ag-enviro_supplement_08-2008.pdf.

Sleeper, Paul. 2008. District Fisheries Biologist, Iowa Department of Natural Resources, Iowa City. Personal communication, November 2008.

U.S. Army Corps of Engineers (Corps), Coralville Reservoir. 2008. "Coralville Lake level information." www.mvr.usace.army.mil/Coralville/waterlevelinfo.htm.

Vitosh, Mark. 2008. District Forester, Iowa Department of Natural Resources, Iowa City. Personal communications, September–November 2008.

Looking Back, Looking Forward

A s long as water has cycled around our planet and fallen to its surface, rivers have risen above their banks to flood surrounding lands. And as long as people have wandered the earth, they have dealt with these rising flows. For eons, human communities have worked to build walls to protect flood-prone areas, or to deepen and straighten river channels, or to hold surging waters behind a dam. But while these defenses may work for small and frequent floods, nature's ancient processes and cycles remain too powerful to control completely. Eventually, an extreme flood overrides the seemingly fool-proof defenses, and people marvel at the destruction it brings. Rivers are meant to flood. The question is not whether more floods will come—they will—but rather when they will rise and how large they will be.

Our task then is to learn to live with floods, maximizing their benefits wherever possible and minimizing the destruction of human constructs. How we do this, and how much we continue to live in harm's way by occupying the river's floodplain, depend on the frequency of flooding, our risk-taking proclivities, and the amount and value of structures and infrastructure we have already established on a given floodplain. Section II described ways in which humans may be pushing the balance toward larger floods. With this in mind, and realizing that even the 2008 floods were not as large as what might someday occur, we need to choose future directions. How will we create a reasonable and wise balance between the present advantages of living near rivers and the inevitable future costs of doing so?

This question has been answered in different ways in different locations. In Venice, Italy, where floods are increasingly frequent, many Venetians have deserted the oft-flooded first floors of their buildings and moved upstairs or out of Venice. This abandonment principle was also adopted in the Mississippi River town of Valmeyer, Illinois, after the 1993 floods: the entire town was moved from the floodplain to a higher location. A contrasting approach involves massive engineering structures to shut floods out. The Japanese, inhabiting mountainous islands where habitable space is limited, have among other things constructed super levees so voluminous they are unlikely to fail. These extremely wide earthworks built alongside lowland urban rivers are large enough to hold multiple buildings several stories high that are constructed on top of the levees (Stalenberg 2007).

The Netherlands, a country with half its land surface below sea level and where many tens of thousands of lives have been lost to floods, depends on large structures for mitigating sea floods. Its Delta Works, a series of dams and storm surge barriers that includes thousands of miles of dikes and hundreds of structures, has been called the world's largest flood protection project and one of the most extensive engineering projects. However, the Dutch are now tackling the threat of Rhine River floods in a different way—through a program called Room for the River. This program involves allowing the rivers more room for floodwaters through reconstructing floodplains, constructing large water retention areas, removing obstructions from floodplains, developing wetlands, and similar activities, rather than continually heightening dikes as floodwaters increase (Fokkens 2007). Closer to home, Mecklenburg County in North Carolina is developing an integrated regional approach to floodplain management. This plan addresses streamwater quality and flood hazards in a coordinated manner, through acquiring floodprone property, protecting floodplains from development, restoring their natural functions, and educating the public. The program has won widespread acceptance because of evidence of improved environmental quality and reduction in flood damage (Hazard Mitigation Section 2003).

These approaches provide examples for consideration. Here in Iowa, where much of our countryside is in agricultural use but remains free of human structures, the range of possible responses to flooding remains large. Here, where we have the open space to do so, modeling our response after the Netherlands' Room for the River program or Mecklenburg County's integrated approach may in most locations be the most effective and least expensive way of minimizing future flood damage.

Section IV considers various responses to the destructive 2008 floods and outlines some techniques for lowering the risk of repeats. Jack Riessen starts us out with a chapter on floodplain management, which is usually tied to managing for a 100-year flood. Herein lies the problem: these events are medium-sized floods, not extreme, and larger floods often surpass the protective restraints of current floodplain regulations (see plate 10). Riessen asks how we can climb out of our "100-year flood silo" to regulate for the inevitable larger floods. He briefly traces the history of floodplain management and outlines ways to positively supplement floodplain regulations.

Douglas M. Johnston follows with chapter 21, which suggests taking a regional approach to floodplain management rather than assuming the standard local flood-remediation efforts. He points out that we will benefit the most by acknowledging that water does not respect political boundaries. Each community's actions affect the hydrology of the entire interconnected watershed. We can work together to maximize the mutual good of all watershed residents, rather than conducting a flood arms race by vetting one community's remediation efforts against those of communities upstream and downstream.

Chapter 22 describes flood barriers, in particular levees. While these may seem like simple structures, authors Nathan C. Young and A. Jacob Odgaard point out that several types exist, that they have both advantages and disadvantages, and that their use highlights the need for well-coordinated regional planning to ensure the greatest benefit throughout the watershed.

Wayne Petersen writes in chapter 23 about specific methods for decreasing urban stormwater runoff. He acknowledges conventional detention basins, but says we need to do more to encourage precipitation to infiltrate into the soil. He goes on to describe ways to increase soil organic matter, the use of bioretention structures, and the advantages of permeable pavement. Use of these infiltration practices will help decrease the common problem of flashiness and flooding in smaller urban creeks.

Chapter 24 tackles the problem of runoff from the row-cropped lands that now dominate Iowa. Laura Jackson and Dennis Keeney explain that in order to minimize flooding and bolster environmental health, we need to find ways to decrease this runoff and to soak up and store more moisture in our soils, tasks that the authors assign to increased perennial vegetation. They describe a number of ways to promote greater perennial cover of Iowa's farmlands.

The section concludes with Gerald E. Galloway's chapter describing the study that he led following the 1993 midwestern floods. This study resulted in a report that is still praised for its broad-ranging conclusions and recommendations. However, few of the recommendations were implemented, and as a result the 2008 floods were in part a repeat of 1993. Galloway writes, "We can no longer afford to ignore the lessons of major floods."

This final book section presents a number of positive methods for addressing and reducing flood damage. Galloway's chapter is a potent proclamation of our duty to future years and to future people. As he concludes, and as all the chapters in this book confirm, "Will we again leave future generations to deal with the challenges that we should have met? We owe it to them to do better." ≋

References Cited

Fokkens, B. 2007. "The Dutch strategy for safety and river flood prevention." In *Extreme Hydrological Events: New Concepts for Security*, ed. O. F. Fasiliev, P. H. A. J. M. van Gelder, et al., pp. 337–352. Netherlands: Springer. For information on the Delta Works, see "Deltawerken Online," www.deltawerken.com/.

Hazard Mitigation Section. 2003. "Integrating water quality into floodplain management—Charlotte–Mecklenburg County's approach." Raleigh, N.C.: North Carolina Division of Emergency Management. http://149.168.212.15/mitigation/case_mecklenburg1.htm.

Stalenberg, B. 2007. "River Basin Management: Symposium Urban Water Management; Japanese state of the art." Delft, Netherlands: TU Delft, Delft University of Technology. Microsoft PowerPoint—*symposium 180107 river basin management*.ppt.

Jack Riessen

20 When (Not If) the Big One Comes

THE IOWA FLOODS OF 2008 once again reminded us of a basic truth: if you live in a floodplain, sooner or later you'll get flooded. And once again the floods of 2008 reminded us that a 100-year flood is just an intermediate-sized flood. Much larger floods can and do occur with some regularity. Homes, businesses, and facilities protected to a 100-year flood level are still vulnerable to flood damages. If we don't climb out of our "100-year flood silo" and plan for much larger floods, 2008 is sure to repeat itself.

By all accounts, the very large 2008 floods greatly exceeded a 100-year flood in a number of locations. The 2008 floods were, however, predictable in that they were well within the realm of what's known to be possible. As high as the floodwaters reached in downtown Cedar Rapids, they were still about 1.5 feet below the level of the Standard Project Flood as shown in a 1967 U.S. Army Corps of Engineers report (Corps 1967). The Standard Project Flood is considered to be the largest flood that can reasonably be anticipated. Despite the inevitability of large-magnitude floods, most floodplain regulations, mitigation policies, and emergency response plans only consider the 100-year and lesser floods.

This nation's early efforts to minimize flood losses attempted to control flooding through the construction of flood control works like dams and levees.

The notion that we instead should learn to live with floods gained traction in the 1950s, when it became apparent that flood losses were continuing to increase despite a significant federal investment in flood control works. The idea was that by mapping the floodplain, regulations could be adopted that either discouraged development in this hazardous area or, at the very least, required buildings and other structures to be flood-proofed. Then, as now, land use regulations were controversial, and an important issue was the size of the flood used to identify the regulatory floodplain. Using the largest flood imaginable would likely have raised constitutional land use challenges by floodplain landowners. On the other hand, use of a smaller flood might lead people to believe that areas outside the regulatory floodplain would never be flooded.

The Tennessee Valley Authority (TVA) wrestled with this issue in the 1950s. The TVA routinely determined the size of two floods for their flood studies, a "maximum probable" flood and a somewhat smaller "regional" flood. It eventually decided to use the regional flood as a regulatory standard due to concerns that using the maximum probable flood would not be acceptable (Wright 2000).

Iowa and a handful of other states that were pioneers in regulating floodplain development also adopted an intermediate-sized flood as their regulatory standard. However, the methods used to size this intermediate flood varied from state to state. Recognizing the need for more uniform methods, a panel of federal agency experts in 1967 recommended the 100-year flood as a standard that would balance the need to reduce flood losses with the need to avoid what might be considered excessive regulation (Wright 2000). The 100-year flood soon thereafter became the de facto standard for regulating floodplain development across the nation.

When the National Flood Insurance Program (NFIP, now administered by FEMA) was created in 1968, it too embraced use of the 100-year standard. Local governments that join the NFIP make federally-backed flood insurance available to their citizens, but in exchange the local governments must agree to regulate development within the identified 100-year floodplain. New houses and businesses within this floodplain, for example, must raise their lowest floor above the 100-year flood level. Existing low-lying buildings are grandfathered in until they are substantially improved or receive substantial damage.

Iowa has had a state-level floodplain regulatory program for more than 40 years and has been the beneficiary of many federal flood control projects like the Coralville, Saylorville, Red Rock, and Rathbun Reservoirs and levee projects in numerous cities. Additionally, most of Iowa's flood-prone cities and many counties have been participating in the NFIP for 20 or more years and have adopted local floodplain regulations. In spite of this, the 2008 floods, the 1993 floods, and a number of other large but less noticed floods have caused significant damage. It's fair to ask: are we doing something wrong? Have floodplain regulations and flood control works failed to curb flood losses?

The answer is no: flood-control works and floodplain regulations have pretty much worked as intended. It's just that a number of floods, as could be expected, have exceeded the 100-year flood. When such big ones come, homes and businesses flood-proofed to the 100-year level are damaged. Some buildings located outside the 100-year floodplain also flood. Water treatment plants with 100-year flood protection quit working, and levees designed to provide 100-year flood protection overtop or fail, suddenly flooding the land and buildings behind them. And when the "big one" comes, the design capacity of flood control reservoirs is quickly used up, and their ability to control downstream flooding is greatly reduced.

The city of Iowa City provides a good case in point. Coralville Reservoir, located upstream of Iowa City on the Iowa River, was constructed in the late 1950s to help control downstream flooding. The city's Planning and Zoning Commission realized that the reservoir did not have sufficient capacity to control all potential floods. Working with the Corps and Iowa Natural Resources Council staff, the commission wisely developed zoning regulations that recognized the city's residual flood risks. These regulations were adopted by the City Council in 1962 after considerable debate (Howe 1969). Two floodplain zones were established: a valley channel zone and a valley plain zone. Only open-space uses were allowed in the valley channel zone, which included the Iowa River channel and a portion of the adjoining floodplain. Buildings were allowed in the valley plain zone provided they were elevated or floodproofed to the regulatory flood level (roughly equivalent to today's 100-year flood level). Iowa City's 1962 floodplain regulations were very similar to the regulatory approach now used by most states and required under the NFIP.

Iowa City joined the NFIP in 1972, and an initial Flood Insurance Study (FIS) was completed in 1976. Based on past studies and reports that considered the flood-control benefits of Coralville Reservoir, the 100-year and 500-year flood peak discharges for the Iowa River in Iowa City were estimated to be 25,000 and 36,500 cubic feet per second (cfs), respectively (FIA 1976). These discharges were used to prepare the flood maps and flood profiles contained in the 1976 FIS, and Iowa City updated its floodplain regulations to incorporate this information.

The 1993 Iowa River flood overwhelmed the storage capacity of Coralville Reservoir, and the peak discharge of the Iowa River reached 28,200 cfs in the downtown Iowa City area, slightly above the estimated 100-year flood discharge at that time (Parrett et al. 1993). Questions about the operation of Coralville Reservoir during the 1993 flood resulted in a re-evaluation of the Iowa River's hydrology and, based on more sophisticated models and the additional years of record since the 1976 FIS, the 100-year and 500-year discharges at this location were revised upward to 29,000 and 45,000 cfs. The flood maps and profiles in the Iowa City FIS were subsequently revised in 2002 to reflect these changes (FEMA 2002), and the city's floodplain regulations were also updated to reflect the new information.

The 2008 floods once again overwhelmed the storage capacity of Coralville Reservoir, and the Iowa River reached a peak in Iowa City of 41,100 cfs, 46 percent larger than the 1993 flood peak (USGS 2009). Buildings that had relatively minor or no damage in 1993 had three, four, or more feet of floodwater in 2008. And a number of buildings lying outside the 100-year floodplain, as shown on the 2002 FIS flood maps, were also flooded.

Everything in Iowa City pretty much worked as had been planned. The Coralville Reservoir did, in fact, prevent a lot of flooding. The 2008 peak discharge of 41,100 cfs was still a lot less than the peaks of 70,000 cfs in 1851 and 51,000 cfs in 1881 (Lara and Eash 1987). Iowa City was a pioneer in developing floodplain regulations and has had relatively good flood maps since the early 1960s. Most buildings built in accordance with floodplain regulations survived the 1993 flood with relatively little damage. By contemporary floodplain management standards, Iowa City has been doing all the right things. The main problem is that the 2008 flood greatly exceeded the 100-year flood standard used for the regulations. No one was asking the questions, "What will happen when the big one inevitably comes?" "Will it be a matter of temporary inconvenience or a major disaster?" "Are we willing to gamble that the 'big one' will not happen any time soon?" And as big as the 2008 Iowa River flood was, it likely was not the "biggest of the big ones" that could potentially occur.

There's no easy answer to whether we should be doing things differently. Some have suggested that a flood larger than the 100-year flood be used as a regulatory standard. But the 100-year flood standard is so ingrained in the NFIP and the resulting flood maps and regulations that changing to a higher standard would take a massive effort and likely meet political resistance. Convincing people to protect buildings and property even to a 100-year flood level is often very difficult. The courts might also invalidate a higher regulatory standard as being overly stringent and unreasonable. There are, however, some other actions that should be considered:

- Require that discharges and flood depths for a Standard Project Flood be developed and mapped for all flood studies. This would provide notice of what is potentially possible.
- Require that critical public facilities like drinking water treatment plants and public record repositories be located outside the Standard Project Flood floodplain or elevated or flood-proofed to this level.
- Better educate the public and professionals in the banking and insurance industries on flood risks. Too often, the evaluation of a building's flood risk is limited to whether it's in the 100-year floodplain. A building marginally outside the 100-year floodplain actually has about a one-in-four chance of being flooded in a 30-year period.
- Determine, and publish, confidence limits for the 100-year and other flood estimates. Flood estimates are just that—estimates. Confidence limits provide a better understanding of the uncertainty involved. For example, it would be more informative to say there's a 90 percent certainty that the true 100-year flood has a peak discharge between 35,000 and 60,000 cfs, rather than just saying it's 45,000 cfs. Local officials might then decide to adopt regulations requiring buildings to be protected to a 60,000 cfs flood level.
- Better educate design professionals on flood risk so that buildings and facilities can be designed to provide additional protection if warranted. In many cases, protection for floods greater than the 100-year flood can be incorporated into a building's design for relatively little cost.
- Better support the Iowa Department of Natural Resources' floodplain management program. This program, in existence since 1949, has never been adequately staffed and funded. A 1998 review showed a program return on investment of $28 to $1 (Towe 1998). But state resources devoted to the program have continued to decline when cost of living increases are considered.
- Take a serious look at flood insurance rates. A common comment is that flood insurance is simply too expensive and, as a result, people don't buy it unless they have to. My analysis, using Cedar Rapids as an example, suggests that flood insurance rates for some buildings are, in fact, much higher than needed to be actuarially sound.
- Start asking "what if?" questions. City planners, emergency managers,

design professionals, and others need to start asking what will happen when the "big one" comes. Will city hall be flooded and records lost? Will water treatment plants cease to function? Will major transportation routes be impassible or, worse, destroyed? Will industries have to shut down for a prolonged period? And what will happen when our 100-year levee is overtopped? If these "what if" questions aren't asked, no one will be prepared when the "big one" does come.

• Work with local governments individually to develop comprehensive flood hazard mitigation plans that consider the full range of potential floods. Each plan needs to be tailored to fit not only local flood risks, but also the local government's long-range goals.

The floods of 2008 reminded us that the 100-year flood is very much an intermediate-sized flood, and that putting all your flood-damage-reduction eggs in a 100-year flood basket is simply not good planning. If we don't climb out of our 100-year flood silo and begin to look more broadly at all floods and flood risks, 2008 is sure to repeat itself at some time, in some place, in Iowa.

References Cited

Federal Emergency Management Agency (FEMA). 2002. *Flood Insurance Study 19103CV000A, Johnson County, Iowa, and Incorporated Areas.* Washington, D.C.: U.S. Department of Homeland Security.

Federal Insurance Administration (FIA). 1976. *Flood Insurance Study, City of Iowa City, Johnson County, Iowa, HUD-FIA-169-318.* Washington, D.C.: U.S. Department of Housing and Urban Development.

Howe, J. W. 1969. "An introductory philosophy of flood plain management." In *Flood Plain Management, Iowa's Experience*, ed. M. D. Dougal, pp. 3–9: Ames: Iowa State University Press.

Lara, O. J., and D. A. Eash. 1987. "Floods in Iowa: Stage and discharge." USGS Open File Report 87-382. http://ia.water.usgs.gov/pubs/iowa.publications.html.

Parrett, C., N. B. Melcher, and R. W. James, Jr. 1993. "Flood discharges in the upper Mississippi River Basin—1993." U.S. Geological Survey Circular 1120-A. http://pubs.usgs.gov/circ/1993/circ1120-a/#pdf.

Towe, L. 1998. "The return on investment of floodplain management at the Iowa Department of Natural Resources." Des Moines: Innovators International, Inc. Available from Iowa Department of Natural Resources, Water Quality Bureau, Des Moines, Iowa.

U.S. Army Corps of Engineers (Corps). 1967. *Flood Plain Information, Cedar River, Linn County, Iowa.* Rock Island District, Rock Island, Ill. Available from the Iowa Department of Natural Resources, Water Quality Bureau, Des Moines, Iowa.

U.S. Geological Survey (USGS). 2009. "Water-Resources Data for the United States, Water Year 2008." U.S. Geological Survey Water-Data Report WDR-US-2008, Iowa sites. http://wdr.water.usgs.gov/wy2008/search.jsp.

Wright, J. M. 2000. "The nation's responses to flood disasters: A historical account," pp. 13–32. A report to the Association of State Floodplain Managers, Madison, Wis. www.floods.org/PDF/hist_fpm.pdf.

Douglas M. Johnston

21 Watershed-Based Flood Management

THE IOWA FLOODS OF 2008 remind us that we live in an interconnected system: the watershed. Each watershed, or drainage basin, collects the runoff (and anything the runoff carries) from a region's land area and feeds it through streams and rivers. Consideration of the entire watershed in flood planning is critical because the areas that flood are determined by watershed conditions upstream of the flood. This complexity makes flood management very challenging and helps explain why flood damages continue to increase. In this chapter, we look at how land-use decisions, local community responses to floods, and other factors can make flooding worse, and how in contrast a watershed-based approach could contribute to better flood outcomes.

To date, most of our planning and response to floods has been local. Individuals and communities attempt to manage flooding through the construction of flood barriers, regulations for floodplain management, and other methods discussed throughout this book. However, the efforts of individuals and communities are limited to the extent of their property or political jurisdiction. Water, as we have repeatedly seen, does not recognize property or political boundaries. So, at any given location, the risk of flooding depends not only on local conditions, but also on distant events and practices. If a community upstream increases its runoff, downstream communities face increased flood risk. And if additional agricultural land is drained, downstream areas face increased risk.

The inability of downstream communities to address upstream sources of flooding creates a defensive strategy. In the face of greater quantities of flood-waters from upstream, the best a downstream community can do is to increase its local protection by building bigger reservoirs or taller levees or investing in greater emergency management and reconstruction resources (for example, expanding and relocating emergency response facilities). Measures to mitigate future flood damages are typically exercised by the local governments that are responsible for post-flood recovery. One result of this local-response emphasis is a flood-protection arms race. Many defensive measures such as levees may protect local land from upstream floodwaters but can in turn pass increased risk of flooding to downstream neighbors, forcing them to respond likewise.

In watershed-based planning, the entire area that contributes to flooding in a region is considered, and flood reduction techniques are strategically applied where they are most effective. For example, upstream land areas may be modified to reduce the volume of water they contribute to the streams and rivers they feed. Instead of a community building flood barriers, it might restore or reconstruct upstream wetlands to hold back rural runoff. Alternatively, flood easements on upstream lands could be purchased, allowing the easement owner to flood that land in time of need, in order to reduce flooding downstream. In this way, a community might purchase the rights to flood relatively inexpensive land in order to reduce flood damages on highly valued land.

Decisions about these types of watershed-based tradeoffs require comprehensive assessment of watershed conditions, including land use, the extent of impervious surface area, urban and agricultural drainage practices, soil properties, past and predicted changes in rainfall patterns (including those associated with climate change), topography, and stream channel geometries. Studies of these and other watershed traits are crucial to estimations of floods and the effectiveness of flood reduction efforts. Watershed-based management also requires comprehensive assessment of risk, land and property values, and community/stakeholder values to decide which strategies are the most cost effective, socially acceptable, and technically or politically feasible.

As complicated as these considerations may appear, we already have a number of tools in place that can serve as models for watershed-based flood regulations. The Clean Water Act and related state and local regulations, for example, effectively establish water quality standards for industrial and municipal pollution sources, and more recently for nonpoint source pollution. While stormwater management is a major focus of these efforts, the regulated elements are pollutants rather than water quantity. Couldn't we establish watershed-based standards for runoff volumes and peak discharges with the goal of reducing floodwaters? Again, some communities now assess stormwater fees to help pay for stormwater management. These fees are linked to the amount of impervious surface on a given property and that property's excess contribution to runoff. If we were to apply the excess contribution standard universally in a watershed, the cumulative impact on flood reduction could be substantial.

Also in our tool bag are techniques that we know will reduce river discharge and flood magnitude. Flood-reducing practices include re-perennialization of agricultural land (see chapter 24), implementation of urban stormwater best-management practices, and other changes to reduce downstream hydrologic stress (Johnston et al. 2006). In some instances, large floodplain areas that have been cut off from a river by levees are acquired, the levees are breached, and floodplain functions (including floodwater storage) are reestablished (Sparks and Braden 2007). In urban areas, reducing soil compaction, increasing permeable paving, and other practices increase local water infiltration and storage to reduce runoff (see chapter 23). As important as urban controls may be, however, river discharges will not change much without reducing the drainage from row-cropped agricultural lands that cover the majority of Iowa. Programs such as the Conservation Reserve Program (CRP) and Wetland Reserve Program (WRP) compensate landowners who put land back into perennial vegetation or restore wetlands, but to alter flood outcomes, these types of efforts must be applied at a much larger scale than they now are. Equally important may be the decision not to put anything vulnerable in floodplains, a practice that we often countermand through making arbitrary changes in floodplain boundaries or the filling of floodplain areas.

Watershed-based planning considers the cumulative impacts of all of these types of possibilities and considerations. It integrates the effects of each flood-reducing effort throughout the watershed, as each community's decisions alter flooding patterns in downstream communities. In application, a watershed-based approach prioritizes the flood mitigation efforts that are most effective across the entire watershed.

Of course, there are many obstacles to watershed-based planning for flood management. One is the distributed authority (and responsibility) of government. A community that builds levees or purchases properties in flood-prone areas is not required to take into account how these actions might affect flood damages in the rest of the watershed. Also, there may be many small communities that, acting individually, cannot implement effective practices, and the difficulties of cooperation make action challenging. Regional or state governmental authority is sometimes used to overcome localized actions, but it is often resisted by local governments and may only add layers of complexity. In addition, our legal system does not establish accountability for downstream flooding impacts. For example, drainage laws do not place any liability on drainage districts for downstream flooding or flood damages (CTRE 2005) because the draining of agricultural land for food and other agricultural production is considered beneficial to the public. However, isn't reduction of flood damages also beneficial? When might flood reduction be more important than agricultural or urban development? These are but a few examples of the social and political challenges faced when flood management is approached locally, rather than in a watershed-focused manner.

Although tremendous political and social changes are required for effective watershed-based management to occur, its advantages and opportunities are well recognized. The *120-Day Report* of the Rebuild Iowa Advisory Commission, for example, recommends that "the state will lead in developing guidance and support for integrated, regional planning to address recovery and leverage multijurisdictional strengths for ongoing initiatives," and "the state will move state policy forward and lead the discussion with regional and local interests on floodplain and watershed management" (RIAC 2008). Some counties and watershed regions have initiated watershed or more comprehensive flood planning efforts. For example, Dane County in Wisconsin adopted a County Mitigation Plan that includes coordination of flood mitigation efforts (Dane County 2004). And researchers have undertaken an assessment of flood storage potential in the Red River watershed of North Dakota (Manale et al. 2006).

In conclusion, floods, even catastrophic floods, are not the extremely rare events we often perceive them to be. There will always be the chance that meteorological odds will fall against Iowa, bringing a rainfall so great that no mitigation, no reservoirs, no levees, and no floodplain management will prevent damage. However, even more certain is the fact that on a regular basis, we will continue to face smaller floods whose damages can be reduced through wise flood mitigation. Because we can expect future flooding, a comprehensive view, based on the physical characteristics of the watershed rather than on political or other boundaries, is needed if Iowa seeks effective and efficient solutions. For watershed-based flood management to work, efforts in data collection, flood assessment, regulatory changes, and political cooperation all must increase. Watershed-based information can be gathered and made available to communities. Laws can be created to facilitate interjurisdictional agreements and accountability. Communities will better plan for the next floods if they are enabled and encouraged to look upon floods as a watershed, not a local, problem.

References Cited

Center for Transportation Research and Education (CTRE). 2005. *Iowa Drainage Law Manual*. Ames, Iowa: Iowa State University, CTRE. www.ctre.iastate.edu/PUBS/drainage_law/drainage_law_manual_complete.pdf.

Dane County. 2004. *Dane County Flood Mitigation Plan*. Madison, Wis.: Dane County Department of Emergency Management. www.danewaters.com/pdf/flood_mitigation/flood_plan.pdf.

Johnston, D. M., J. B. Braden, and T. H. Price. 2006. "Downstream economic benefits of conservation development." *Journal of Water Resources Planning and Management* 132(1): 35–43.

Manale, A. P., S. Hanson, and B. Bolles. 2006 "Waffles are not just for breakfast anymore: The economics of mitigating flood risks through temporary water storage on agricultural land in the Red River Basin." *Journal of Soil and Water Conservation* 61(2): 52A(6).

Rebuild Iowa Advisory Commission (RIAC). 2008. *120-Day Report to Governor Chet Culver*. Des Moines: Rebuild Iowa Advisory Commission. www.rio.iowa.gov/assets/RIO_120_DAY_REPORT.pdf.

Sparks, R., and J. Braden. 2007. "Naturalization of developed floodplains: An integrated analysis." *Journal of Contemporary Water Research and Education* 136: 7–16.

Nathan C. Young
A. Jacob Odgaard

22 Flood Barriers

~~~~~~

FLOOD BARRIERS HAVE been heavily featured in news stories and discussions involving the 2008 floods on the Iowa and Cedar Rivers. Permanent and temporary flood barriers were used to combat flood damage during the event. But in many cases, these barriers were overtopped or failed, causing significant damage to public and private property. Following the event, flood barriers became major components of proposed flood mitigation plans. However, while flood barriers can be effective in protecting lowland areas from flooding, the financial, ecological, and social costs associated with them can be prohibitive. As Iowans recover from the flood and seek protection from future disasters, it is important for the public and policy makers to understand the capabilities, value, and limitations of flood barriers in reducing flood risks.

Flood barriers are permanent or temporary structures designed to contain floodwaters and prevent them from inundating adjacent low-lying areas.[1] Several types of flood barriers are depicted in figure 22-1. The most common permanent structures are earthen levees. Levees are large soil embankments constructed along the bank of a river. Levees extend across a floodplain, connecting areas of higher elevation. Levees feature gently sloping sides to ensure soil stability and create mass sufficient to hold back floodwaters.

In urban environments and elsewhere where property is valuable and space is limited, construction of levees may not be feasible. A second type of permanent flood barrier, the flood wall, may be used in these areas. Flood walls, like levees, are constructed next to rivers and streams. They are vertical walls constructed from metal, concrete, or masonry, which are anchored into the ground and commonly feature structural supports to withstand the pressure of floodwaters.

Where permanent structures are not feasible or justified, temporary emergency barriers may be built at the time of flooding. Sandbag levees were extensively used during the 2008 floods. These consist of sand-filled burlap or plastic bags stacked in various geometries to withstand the pressure of anticipated flood depths. They are typically covered in plastic sheeting to decrease leakage. Temporary barriers can also be constructed from soil, rock, and human-made

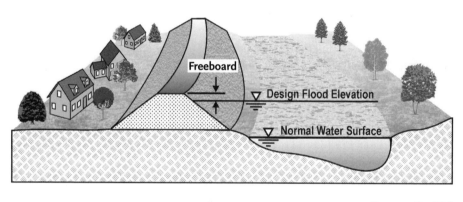

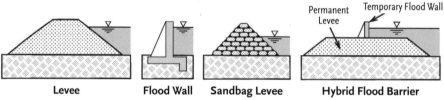

FIGURE 22-1 Flood barriers are constructed along river channels to prevent floodwaters from affecting adjacent property. Levees and flood walls are permanent flood barriers. They are typically designed to protect against 100-year flood elevations plus an additional height, called freeboard. Freeboard, typically 1–3 feet, accounts for the possible effects of waves, settlement of the embankment, and error in flood elevation estimates. Temporary flood barriers, such as sandbag levees, are constructed when floods threaten and are used in areas where permanent barriers are not feasible. Hybrid flood barriers combine permanent and temporary components to reduce cost and maintain community connectivity with the river. *Illustration by Michael Kundert.*

materials. Temporary barriers are versatile and can be relocated or augmented when necessary to adjust to rapidly changing flood conditions.

Hybrid flood barriers, which have both permanent and temporary components, can be useful in providing flood protection in urban areas without necessitating high levees or flood walls that cut off a community's connection to the river. Hybrid barriers feature low permanent levees constructed to protect against more frequent, less severe flood events. These levees eliminate the need to regularly construct temporary flood barriers for small floods, while maintaining access to and even enhancing enjoyment of the river by facilitating construction of walkways and trails. When more severe floods threaten, protection is augmented by placement of temporary sandbag levees or removable flood walls on top of the lower permanent structures.

Flood barriers can significantly decrease the risk of flood damage. However, contrary to common thought, they are not intended to prevent all flooding at all times. Foolproof protection by flood barriers is neither possible, nor is it their intent. For one thing, flood barriers must be carefully designed and maintained to ensure they do not fail or create flood problems for the areas they are intended to protect. One significant consideration is interior flooding. Interior flooding occurs when flood barriers prevent local runoff originating from the barrier's landward side from draining to the river. To prevent ponding of water on the landward side of a levee or flood wall, land is reshaped and tiled to drain local runoff to a central location. It is then conveyed through or over the barrier by pumping or gravity drainage through pipes. Check valves are installed to ensure that rising floodwaters do not flow backwards into the protected area behind the levee.

Improper design or maintenance can cause earthen levees to fail. High velocity water or wave action can cause erosion of the riverward levee slope. Sometimes vegetation can protect levees from erosion. Large rock (riprap), concrete mats, or pavement also can armor the slope. Levees may fail due to piping, the development of pathways in or beneath the embankment through which water flows. These pathways grow larger as the moving water scours away the soil, and they can result in a levee breach. Clearing deep-rooted vegetation and controlling seepage can prevent piping. Low permeability clay, steel sheet pile, or concrete curtains are used to prevent piping by impeding movement of water through the soil. Drainage systems and relief wells also reduce piping by relieving upward water pressure from beneath the embankment.

In addition to concerns about design and maintenance, there are inherent limitations to the protection that levees can provide. Financial and spatial concerns prevent the construction of levees capable of containing the largest conceivable flood. Instead, levees are designed to protect against smaller floods, most commonly the 100-year flood (a flood with a 1 percent chance of occurring in any given year). Levee design is based on analysis of flow rates, water surface elevations, and historical flow measurements, using a combination of statistical and computational techniques.

While flood barriers can decrease the risk of flooding along the sites of their construction, they may actually worsen both upstream and downstream flood conditions. Levees confine floodwaters to areas immediately adjacent to the river channel, preventing them from spreading out across their floodplains and slowing in speed (as floodwaters naturally do when they have free access to broader floodplains). When levees remove these natural attenuating effects of floodplains, floodwaters increase in depth upstream and are transmitted downstream more quickly. As a result, flood peaks may increase in magnitude both upstream and downstream of the flood barrier.

Flood barriers may produce several other negative consequences, which must be weighed against their potential benefits. They can, for example, degrade riverine ecosystems by physically altering the river channel and reducing the connectivity between the river and its floodplain. When flood flows are confined by levees, flow velocities are increased along the protected reach. Higher velocities can scour the riverbed, sweeping away fine sediments and reshaping the river cross-section, making it more uniform. These changes decrease in-stream habitat diversity, affecting aquatic species and altering ecological processes.

Barriers also can reduce environmental connectivity between a river and its floodplain and thus limit the natural exchange of sediments, nutrients, and organic matter. This affects terrestrial habitat diversity, food availability, and natural processing of nutrients and other chemicals. Flood barriers also prevent annual flooding of areas important to the life history of many plant and animal species. Some species are adapted to exist in areas that experience annual flood pulses. When floodwaters are excluded from the floodplain, these species lose their competitive advantage over less flood tolerant species.

In addition to potential environmental impacts, permanent flood barriers can have negative social impacts. They disconnect communities from the river by limiting recreational access and obstructing views of the river's natural beauty.

Temporary flood barriers, especially sandbag levees, may also have unintended consequences, including negatively affecting other mitigation efforts upstream and downstream. This problem is especially difficult to foresee and address when individuals, neighborhoods, and communities are mobilizing to construct their own temporary flood barriers in response to an impending flood. In such situations, time, expertise, and resources to properly plan and construct a successful temporary flood barrier may not be available.

Perhaps the most severe unintended consequence of flood barriers is the false sense of security they create regarding elimination of all future floods. This can result in the development of high-value floodplain communities and a lack of preparedness for extreme floods. This was especially true in 2008 in Cedar Rapids, a large city occupying an expansive floodplain. Once the Cedar River's level rose above the permanent and temporary levees that laced downtown Cedar Rapids, the floodwaters quickly spread to inundate the entire floodplain area. This shocked many business and homeowners who never imagined that floodwaters could reach their properties, which were previously believed to be protected by the flood barriers.

The many broader impacts of flood barriers, described above, highlight the necessity for well-coordinated regional flood mitigation programs, or integrated watershed partnerships. A successful partnership would bring together individuals, communities, and government agencies to coordinate flood mitigation efforts on a basin-wide scale. The partnership would provide an avenue to facilitate the sharing of knowledge, from professional engineering expertise to a community's knowledge of past flooding events and challenges. Members of an integrated watershed partnership would work as a team to establish mitigation priorities. In addition, the partnership would naturally help generate community support for proposed mitigation strategies, including flood barriers. In coming years, we are likely to have detailed computer-based hydrologic models that will be used by citizens and professionals to understand flood flows and their mitigation from an integrated watershed perspective.

In conclusion, as Iowa communities develop strategies to address flood risks, it is important that the strengths and limitations of individual flood mitigation tools be carefully considered for each site along a given river. In addition to flood barriers, other mitigation alternatives will likely be needed along eastern Iowa's river corridors, with mitigation efforts being fitted to the individual needs of each location and each stakeholder group. Thus, broad-scale cooperation among individuals, communities, and government agencies is needed to ensure that individual flood-control projects are complementary to one another at the basin scale, and that they provide the greatest benefit to everyone. In addition, it is important to educate those who live near flood barriers so that they truly understand the risk of overtopping by a flood event that exceeds the barrier's design standards.

Note

1. Information in this chapter is based on USCFR (2008), Corps (2000), and FEMA (1986).

References Cited

Federal Emergency Management Agency (FEMA). 1986. *Floodproofing of Non-Residential Structures.* Washington, D.C.: United States Department of Homeland Security. www.fema.gov/library/viewRecord.do?id=1413.

U.S. Army Corps of Engineers (Corps). 2000. *Engineer Manual 1110-2-1913, Engineering and Design—Design and Construction of Levees.* Washington, D.C.: United States Department of the Army. http://140.194.76.129/publications/ eng-manuals/em1110-2-1913/toc.htm.

U.S. Code of Federal Regulations (USCFR). 2008. *Mapping of Areas Protected by Levee Systems.* United States Code of Federal Regulations Title 44, Section 65.10. http:// edocket.access.gpo.gov/cfr_2008/octqtr/pdf/44cfr65.10.pdf.

Wayne Petersen

23 Managing Urban Runoff

As I DESCRIBED IN chapter 9, urban landscapes and new urban development increase the volume of stormwater runoff by expanding the proportion of land covered with impervious surfaces. This increased urban runoff certainly contributed to Iowa's 2008 flooding. But because of Iowa's relatively small total coverage by urban areas, urban runoff could not have been a major contributor to these massive floods.

However, urbanized landscapes periodically cause flooding in smaller urban stream corridors. This flooding is the number one natural hazard facing Iowa's urban communities. The hazard rises as a watershed's imperviousness increases. Therefore, managing urban runoff more sustainably is a justifiable need that should be addressed for flood control, as well as for water quality protection.

Managing urban stormwater runoff is not a new concept. Many communities have had ordinances for years that require stormwater management. These ordinances focus on flood control. They typically require installation of detention basins in new developments. Detention basins are depressions or dammed areas designed to catch and hold runoff from large storms. Detention basins reduce downstream flooding by controlling the rate at which impounded water is released. However, detention basins do not lessen the total quantity of urban runoff. They simply slow its release, discharging it over a longer period of time. Detention basins may reduce peak flows, but they do not reduce the increased volume of urban runoff associated with new development.

Another shortcoming of detention basins is that they only manage runoff from large rainstorms. Most stormwater ordinances require detention of rainfall events of about 4 to 7 inches per 24 hours. These are 100-year to 5-year storms, which have a 1 percent to 20 percent chance of happening in a given year. Runoff from rains of less than 4 inches generally moves through detention basins unmanaged. These smaller storms can still cause flash flooding of urban streams, especially if a number of such rainfall events occur in succession (Foreman 2008).

To reduce runoff and flooding in watersheds undergoing urban growth, we must mitigate the increased runoff generated from the impervious and compacted surfaces associated with new development. We also must retrofit older urban landscapes, which were designed to shed urban runoff as rapidly and efficiently as possible. While such retrofitting is challenging and costly, it is necessary if we are to create stronger, more sustainable communities in the future. We can address both new and older urban areas by using infiltration-based stormwater management practices, which are being increasingly utilized across the country and incorporated into the design of new developments. These relatively new practices are the topic of the remainder of this chapter.

The new stormwater management practices attempt to restore or mimic the infiltration-based and groundwater-driven hydrology that once existed in Iowa by eliminating runoff from smaller rains that account for the majority of annual precipitation (figure 9-2). While these new management practices are directed primarily toward improving water quality, they also reduce stream flashiness and the magnitude and duration of floods in smaller urban watersheds. Detention basins most likely will continue to be required to provide flood control when large storms occur. There are examples though where infiltration-based practices have been designed to handle the large storms, and detention basin requirements have been waived.

Infiltration-based practices focus on reducing runoff from Iowa's small, frequent rains because these rains transport the majority of pollutants to receiving waters. An analysis of Iowa rainfall data from across the state has determined that in 90 percent of past rains, the accumulation has been less than 1.25 inches in 24 hours (CTRE 2008). Due primarily to the impervious surfaces and compacted soils in urban areas, a portion of the rain from a 1.25 inch storm would run off. This amount of water has been designated as the water quality volume (WQv). It is an amount that must be captured and stored to mitigate water quality impacts. However, the best strategy for treating the WQv would be to infiltrate this water (or as much of it as possible) within the urban landscape itself.

The following paragraphs discuss three of several infiltration-based practices in the *Iowa Stormwater Management Manual* (CTRE 2008): soil quality restoration, permeable pavement, and bio-retention.

Soil quality degradation is a major cause of our landscape's reduced hydrologic capacity. Conversely, soil quality restoration offers our best chance for reducing runoff in the shortest time, on the most comprehensive scale, and with the least cost. Therefore restoring soil quality needs to be a key strategy for helping landscapes hold and infiltrate more rainfall.

This is obviously true of the agricultural lands that dominate Iowa.[1] But soil quality restoration is also necessary in urban areas. New development always involves land-disturbing activity that alters and compacts soil. Instead of creating compacted green space that generates runoff, we need to create lawns that can infiltrate runoff from nearby impervious surfaces.

Urban soil-quality restoration can be performed during final grading and landscaping. It involves deep tillage to break up compacted soils and amending soils with compost to enhance organic matter content and increase soil porosity. Deep tillage shatters compacted soils and provides channels for water to move down into the soil profile. The high organic matter content of compost helps the landscape absorb rainfall. Applying a 2- to 3-inch blanket of compost and tilling to a depth of 8 inches yields an organic matter content of 5 percent to 8 percent and provides about 40 percent soil pore space. Such soils should infiltrate the WQv (CTRE 2008; Soils for Salmon 2008). On existing turf, annually applying a half-inch blanket of compost after aeration will build a lawn's capacity to absorb more rain and shed less runoff (figure 23-1). Some landscaping and lawn care businesses are beginning to offer soil quality restoration services. The city of Ames now requires soil quality restoration in new public development projects.

Permeable pavement, our second major infiltration technology, can significantly reduce runoff. Studies have shown that in urban areas, 60 percent to 70 percent of imperviousness is related to transportation surfaces (e.g., roads, parking lots, driveways) (EPA 2005).

There are a number of permeable pavement alternatives. Porous asphalt

FIGURE 23-1 Rather than laying sod, this developer is applying a compost blanket with seed added to the top quarter inch of the compost. Compost seedings provide rapid germination and establishment of turf. The resulting lawn absorbs more rain and requires less watering because of the high organic matter content of the compost. Note the dark color of the organic-rich compost. *Photograph provided by the U.S. Natural Resources Conservation Service.*

and pervious concrete bind together uniformly sized aggregate (rock particles) to create a pavement that resembles a rice krispie bar. Water moves down through the pore space in the pavement. Permeable paver blocks have spacers that prevent one block from lying tightly against another one. Small granules of rock fill the void space between paver blocks.

Regardless of the type of permeable pavement used, the system requires a bed of clean, angular aggregate beneath the pavement for bearing strength and for temporary water storage in the pore space between the aggregate (figure 23-2). The below-ground rock chamber can be designed to handle any size rainfall event—from a 1.25-inch event to a 100-year storm (about 7 inches of rain in 24 hours). Designing permeable pavement systems to manage the 100-year storm eliminates the need for detention basins, which can occupy highly valuable developable land.

Water that percolates through permeable pavement is cleaned, cooled, and slowly released. A biofilm develops on the aggregate in the rock chamber. The biofilm hosts microbes that capture and break down pollutants. Infiltrated rainfall moves out into the surrounding soil and becomes groundwater flow, if it can. If surrounding soils won't allow water to percolate, a tile (perforated pipe) installed in the rock chamber will allow clean, cool water to be released in a manner that mimics groundwater discharge. Both water quality and flooding problems can be addressed with permeable pavement.

Because permeable pavement is relatively new in Iowa, we do not have long-term examples of its performance. But these systems have proven themselves

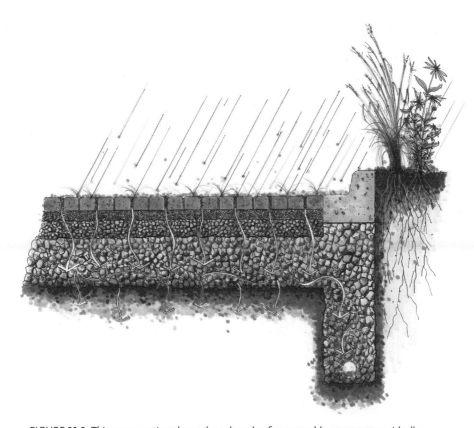

FIGURE 23-2 This cross-section shows the subgrade of a permeable paver system. Ideally, water that moves down through the pavement and rock chamber will percolate into the surrounding soil and become groundwater flow. If it can't move into the soil profile, water will move to a subdrain tile and be discharged. Water percolating through the rock and soil will be cleaned and cooled and slowly released, rather than flowing as hot, dirty water through a storm sewer into a receiving stream. *Illustration provided by the U.S. Natural Resources Conservation Service.*

to be successful for a number of years in other cold-weather locations. Proper design, installation, and maintenance are key to ensuring permeable pavement's success. Annual vacuuming (with a vacuum truck for large areas, or other devices for smaller installations) is instrumental to ensure that particulate matter does not accumulate and plug pore space on the pavement surface.

The third infiltration-based practice, bioretention, involves creation of shallow bowls or depressions that are strategically located to accept runoff from impervious surfaces (figure 23-3). The runoff is temporarily impounded and then infiltrated through a mixture of sand, compost, and topsoil.

Rain gardens are a bioretention practice that is growing in popularity. Rain gardens are depressional gardens that typically take runoff from the roof of a house, but some accept runoff from streets. A rain garden relies on a healthy soil profile that has good natural percolation rates (i.e., water moves down through the soil at the rate of at least 0.5 inch per hour) (Petersen et al. 2008).

After new construction, soils are typically altered and compacted so that an engineered subgrade is required to adequately percolate impounded water. As with permeable pavement systems, the soil matrix of a bioretention cell hosts microbial populations that capture and break down pollutants. Infiltrated water either becomes groundwater flow, or it is slowly released through a tile system after it has been cleaned and cooled.

Typically bioretention practices impound water to a depth of 6 to 9 inches. The engineered subgrade includes a 12-inch gravel bed at frost line (42 to 48 inches down). A perforated drain tile is placed in the gravel bed. A 24-to-30-inch-deep soil matrix covers the rock chamber. The soil mix consists of about 60 percent washed concrete sand, 25 percent compost, and 15 percent topsoil. This sandy loam should percolate about 1 inch of water per hour, allowing impounded water to drain down in 12 to 24 hours (CTRE 2008).

Bioretention cells can be vegetated in a variety of ways. Deep-rooted native plants are encouraged. These plants withstand a range of climatic conditions, contribute abundant organic matter to the soil matrix over time, and maintain high porosity to ensure continued adequate percolation.

What can we do to hasten the adoption of infiltration-based stormwater management? As of December 2008, 43 cities in Iowa were required by Clean Water Act regulations to pass ordinances that deal with stormwater management from new construction and to protect water quality to "the maximum extent practicable." In my opinion, the Environmental Protection Agency should add definition to these ordinances by requiring management of the WQv through infiltration. This would provide direction to municipalities required to draft new

FIGURE 23-3 This bioretention cell (often referred to as a rain garden) manages street runoff. A curb cut allows street runoff to flow into the bio-cell. This was installed by a landscaping business (Forever Green Nursery, Coralville, Iowa) offering sustainable landscaping services, a new growth market for the industry. *Photograph by the author.*

stormwater ordinances. It would also provide consistency from one municipality to another and tangible water quality benefits across Iowa and the nation.

I also think we should reconsider the design standards of these infiltration systems. As mentioned above, design standards now handle runoff from the current WQv for Iowa. In light of the 2008 flooding and the tendency toward larger storms, we might want to revisit design standards for infiltration-based practices and design them for larger rainfalls. Perhaps we should design to infiltrate the 1-year storm (the amount equaled or exceeded about once a year), which in Iowa is an average of about 2.4 inches of rain in 24 hours. Or we could infiltrate the 2-year storm (the amount equaled or exceeded about once every 2 years, or about 3 inches of rainfall in 24 hours), or even the 5-year storm (the amount equaled or exceeded about once every 5 years, or just under 4 inches of rainfall in 24 hours). Detention basins could still be used to impound runoff and control release rates of larger storms but hopefully could be downsized if more rainfall is infiltrated (CTRE 2008).

Raising design standards would be a challenge, even though doing so would protect water quality and reduce flood potentials. Much of Iowa has yet to embrace infiltration-based stormwater management, even to prevent runoff from 1.25 inches of rain. But the floods of 2008 offer opportunities to think differently about future needs.

In closing, I want to reinforce two recommendations the Iowa Department of Agriculture and Land Stewardship (IDALS) made to the Rebuild Iowa Task Force on Agriculture and the Environment (Perkins 2008). IDALS recommended new initiatives for infiltration and water storage projects. These initiatives encompass the sustainable water management practices outlined in this chapter. While they are directed toward improving water quality and mitigating smaller floods, they will also reduce the runoff associated with very large floods such as those of 2008. As such, these water management practices can become one part of a more comprehensive, multifaceted plan to mitigate future major flooding in Iowa.

IDALS also recommended that all Iowa residents, especially those living in watersheds that flooded in 2008, think about the hydrologic footprint of their own activities. We can ask ourselves these questions: Did it rain on my property in 2008? If so, did my property generate runoff? If the answer is yes, as it almost certainly will be, then ask what you can and will do to reduce the volume of runoff you will contribute to future floods. Consider what infiltration-based practices you can adopt to store more water on your own land. By doing so, you will be taking a small but significant step toward reducing damages of future floods and toward building hydrologically sustainable communities for generations to come.

Note

1. A short paper on soil quality restoration needs and recommendations for agricultural land is available from the author. Contact Wayne Petersen at wayne.petersen@iowaagriculture.gov or 515-281-5833.

References Cited

Foreman, Laurel. 2008. Hydrologist, Natural Resources Conservation Service, Des Moines, Iowa. Personal communication, December 2008.

Center for Transportation Research and Education (CTRE). 2008. *Iowa Stormwater Management Manual.* Ames, Iowa: Iowa State University, CTRE. www.ctre.iastate.edu/PUBS/stormwater/index.cfm.

Environmental Protection Agency (EPA). 2005. *Management Measures to Control Nonpoint Pollution from Urban Areas*, publication EPA 841-B-05-004. Washington, D.C.: EPA. http://epa.gov/nps/urbanmm/pdf/urban_guidance.pdf.

Perkins, J. 2008. "Soil, water conservation took a beating in floods." *Des Moines Register* July 30, 2008.

Petersen, W., et al. 2008. *Rain Gardens: Iowa Rain Garden Design and Installation Manual.* Ankeny, Iowa: Iowa Stormwater Partnership. www.iowastormwater.org (search for *Rain Garden Manual*).

Soils for Salmon. 2008. www.soilsforsalmon.org, accessed December 2008.

Laura Jackson
Dennis Keeney

24 Perennial Farming Systems That Resist Flooding

A DROP OF RAIN that falls on Iowa has a 63 percent chance of falling on a corn or soybean field (NASS 2007). If we look just at northern Iowa, where farming is most intensive, that probability rises to 88 percent (figure 24-1). Is this droplet likely to contribute to flooding? To answer this question, let's trace the course of three hypothetical raindrops.

The first drop, falling in summer when crops are actively growing, soaks into the earth, is absorbed through crop roots, and evaporates back out into the atmosphere through crop leaves (a process called transpiration).

The second drop, falling during the 9 or 10 months when growing crops are absent or still very small, soaks into the earth and finds a temporary home in the spongy network of tiny pores, root channels, earthworm tunnels, and ancient prairie humus we call soil. But that raindrop soon enters the nearest ditch or stream, either by entering a drainage tile line or by seeping through the soil and immediately back into the stream.

The third drop runs off the ground surface because it can't soak into the earth. Runoff happens naturally when soil is saturated or frozen. Soils damaged by heavy farm equipment, continuous corn production, or soil erosion are more prone to shed water.

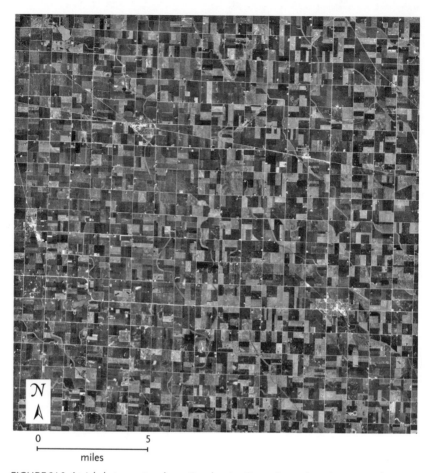

FIGURE 24-1 Aerial photo, centered over Pocahontas County in northern Iowa. Note the dominance of rectangular fields of row crops, which cover most of the ground. In many Corn Belt counties, row crops occupy close to nine-tenths of the land surface. *Illustration from Iowa State University Geographic Information Systems Support and Research Facility. 2006 Orthophotos—USDA (natural color) of Pocahontas County, Iowa. Retrieved May 8, 2007, from Iowa State University Geographic Map Server, www.cairo.gis.iastate.edu.*

Considering these three possibilities, we begin to see the basic principles of creating a flood-resistant landscape, principles as timeless and predictable as gravity. To prevent rains from flowing rapidly into channels and raising water levels, some of their moisture must be returned directly to the atmosphere or discharged steadily and slowly (not in flashy gushes) into drainage ways. Agriculture can help achieve these ends and thus make the land more flood-resilient if it can:

1. Minimize runoff by increasing the speed at which water soaks into the soil and the quantity of water the soil can hold.
2. Store an abundance of water in healthy soils that are high in organic matter, instead of immediately draining it into streams.
3. Increase the amount of time that growing crops are pumping water back into the atmosphere.
4. Intercept any runoff from intense rainfall, passing it into nearby wetlands or through buffer strips that protect streams.

The truth is, the corn and soybeans that now dominate Iowa's agriculture do not help do any of these things. It would be difficult to find two crops that do a worse job of handling Iowa's rainfall. In conjunction, these croplands pollute our waterways by their release of sediment, fertilizers, and pesticides. Scientists studying the problems of surface and groundwater contamination, the Dead Zone in the Gulf of Mexico, and flooding have arrived at the same conclusion: we need to re-perennialize the landscape. Quite literally, we have to rediscover and cultivate our deepest roots.

Before the coming of European Americans in the mid-1800s, the prairie soils of Iowa were filled with a dense and deep underground network of perennial plant roots (figure 24-2). These roots filled the soil year round, not just in July and August. They made the soil more crumbly, porous, and spongy. They shielded it from the destructive power of raindrops. Year in and year out, perennial plant roots added humus to the soil, built pores, and supported a rich community of arthropods, fungi, and bacteria. The perennial sod could absorb tremendous quantities of rain without producing runoff. Perennials active in early spring began using water in April, and the diversity of plant growth ensured that soil water was used through October. Excess water not used by plants steadily percolated into the shallow groundwater and onto wetlands and streams, or it recharged deeper aquifers. Many features made this landscape flood-resistant. It is no wonder that Knox (2006) describes the agricultural conversion of prairie and forest in the upper Mississippi River basin as "the most important environmental change that influenced fluvial [river and stream] activity in this region during the last 10,000 years."

It is tempting to lay the job of restoring perennial plant cover at the feet of our public lands, existing restored prairies, and farm set-aside acres such as those enrolled in the Conservation Reserve Program (CRP). These lands are tremendously important. They provide numerous benefits, including slowing their share of raindrops. But because so much of Iowa is cropland, agriculture

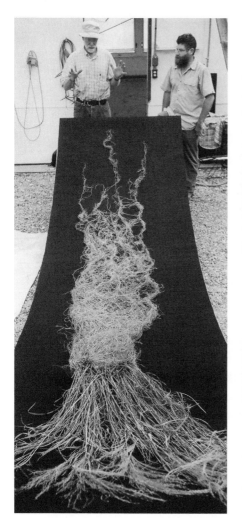

FIGURE 24-2 *Left*: Roots of perennial prairie plants can extend many feet into the soil. Roots and their associated microflora take up nutrients and water and store carbon captured during plant photosynthesis in the soil. *Photograph by Scott Bontz, the Land Institute.*
Right: Close-up of roots of several perennial prairie grasses and forbs, showing dense structure. This root mass armors the soil against the forces of water and wind and quickly captures nutrients before they can leach below the root zone. Plant roots release carbon into the soil and support the activity of beneficial microorganisms, fungi, and arthropods. All this activity builds humus (organic matter) in the soil, which can absorb more water and hold it longer than can simple mineral soil alone. Because the roots of prairie plants grew over thousands of years, Iowa soils have some of the highest levels of organic matter in the world. Plants shown here were grown for two years in a 24-inch-wide polyvinylchloride pipe buried vertically. The pipe was filled with a crushed clay growth medium. The roots were preserved after extraction using fine sprays of water. *Photograph by Laura Jackson.*

must play a major part in restoring hydrological resilience. Encouraging this likely will mean compensating landowners for the many benefits they provide through their land use practices: food of course, but also water quality, wildlife habitat, carbon sequestration, and energy from biomass, in addition to hydrological resilience.

Agriculture can increase perennial plant cover and contribute to Iowa's hydrological health in a variety of ways, some of which are tested and familiar: Corn and soybeans could become part of four- to five-year crop rotations that include small grains, hay, and pasture. Long crop rotations were in wide use up until the late 1950s (figure 24-3) (Jackson 2002), when row crops covered about half of Iowa's planted acres,[1] instead of the current 92 percent (NASS 2007). Today we can improve on the farming of the 1950s by implementing conservation tillage, better options for pest control, intensively managed rotational grazing,

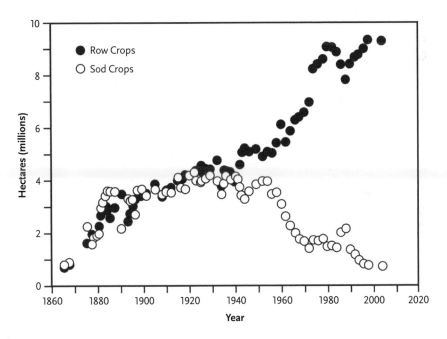

FIGURE 24-3 History of cropping in Iowa, 1860 to 2000. All crop acreages were lumped into two categories: row crops and sod crops. Row crops are grown in widely spaced rows and include mainly corn and soybeans. Sod crops include perennial vegetation such as hay and densely sown small grains including oats, wheat, barley, triticale, and rye, usually undersown with pasture grasses and legumes. Permanent pasture is not included in these statistics. Iowa farmers maintained about 50 percent sod crops for about 100 years, until the late 1950s. Then, although corn acreage remained fairly stable, soybeans began to replace sod crops and greatly increased the proportion of land in row crop coverage. *Adapted from Jackson 2002.*

and cover crops to protect the soil before the row crops are well established. Long crop rotations have become economically feasible once again with the price premiums now offered for organically certified grains and meat.

A more flood-resistant landscape would feature more grass-based farming; thus livestock would play a larger role. Health-conscious consumers are currently turning to leaner pasture-raised beef, dairy, and poultry. Integrating crops and livestock on farms can enhance economic opportunities in rural areas because of the associated need for more labor, barns, fences, water tanks, veterinary services, etc.

Back out in the corn field, we need to critically examine subsurface water (tile) drainage. Farmlands are tiled to dry out the root zone as fast as possible. This has caused more water to flow into our rivers in the spring, instead of lingering in the soils (Schilling and Helmers 2008). Thus, as happened in 2008, rivers are often running full when a really big early summer storm arrives.

Researchers at Iowa State University are modifying strip-cropping practices (which now alternate row crops with European pasture grasses or legumes, such as alfalfa, in strips along the contour) by using deeper-rooted native prairie strips instead. Preliminary results indicate that prairie strips covering just 10 to 20 percent of the total field area were able to reduce sediment loss by 95 percent (losses of 8.5 tons/acre were reduced to 0.5) (Helmers 2009). Prairie strips should also be able to draw down soil moisture earlier in the spring and later in the fall. By occasionally moving these strips, better soil structure could be restored throughout a field.

Taking the use of perennials one step further, they could be incorporated directly into grain production. Farmers in Australia have pioneered a new method to grow their winter grains (oats, wheat, and barley) in the same field with perennial warm-season pasture grasses. And for over 25 years, researchers at The Land Institute in Kansas have worked to develop high-yielding mixtures of perennial grains (Cox et al. 2006). Recently they have made rapid progress by hybridizing wheat, sorghum, and sunflower with their wild perennial relatives. Perennial grains could revolutionize our whole way of doing agriculture, but a great deal more research will be required to develop crops with economic yields.

In 2007, Iowa had the industrial capacity to convert 30 percent of its corn into ethanol (Iowa Corn Promotion Board 2009). If we want to use our productive soils for fuel instead of food, then why grow corn? Mixtures of native, perennial prairie plants could produce as much energy per acre as corn-based ethanol, once the energy needed to produce corn is factored in, while also providing wildlife habitat, regulating water, and storing carbon in the soil (Tilman et al. 2006). The Iowa legislature has supported research at the University of Northern Iowa to test this idea at a large scale (figure 24-4) (Smith 2008), and similar efforts are moving forward in neighboring states. There are many questions about which technologies, scale, end use, and forms of biofuel energy (solid, liquid, electric) would be most practical.

We have solid evidence that perennialization moderates flooding. In the 1930s, the Coon River watershed in southwestern Wisconsin was the site of a basin-wide demonstration project to reduce the soil erosion and flooding that had plagued the area. Half of the 800 farmers in the 90,000-acre watershed participated in a wide variety of conservation practices on three-fourths of the farmland, including the conversion of crop fields to perennial pastures. Runoff volumes and flood peaks were significantly decreased, by as much as

FIGURE 24-4 University of Northern Iowa graduate students prepare to measure biomass productivity from prairie grasses in order to estimate energy availability. The Tallgrass Prairie Center's Prairie Power Project is investigating ways to sustainably harvest prairie biomass while still providing wildlife habitat. *Photograph by Laura Jackson.*

50 percent for extreme events (Krug 1996). Today, descendants of those same farmers continue to protect the valley from erosion and flooding by farming on the contour, using perennial grasses and alfalfa in their crop rotations, and restricting row crops to gently sloping soils (Hawkins 2002).

Although cropland runoff can be decreased, some will remain. This should flow into a wetland or buffer strip before reaching a stream (figure 24-5). (Keep in mind that most wetlands were originally fed by groundwater, and never functioned as repositories for heavy runoff, so constructed wetlands are something of a bandage solution.) Buffer strips, 66 feet wide, of prairie grasses, shrubs,

FIGURE 24-5 A buffer strip of mixed herbaceous and woody perennial vegetation along Bear Creek in central Iowa. The buffer intercepts surface runoff and sediments before they can enter the stream. *Photo by Lynn Betts, U.S. Natural Resources Conservation Service.*

and trees along streams will trap sediments, improve soil quality, remove some nitrogen and phosphorus, and increase water infiltration into the soil. By breaking selected field tiles as they meet the buffer strip, we could further slow the movement of water into streams.

For years, we have mistaken wetlands as idle. In reality, a wetland is like a baseball field in winter, standing ready to do its job when the time comes. Mitsch and Gosselink (2000) estimated that to minimize the next large-scale flood on the Mississippi River, we would need about 7 percent of the river's basin in wetlands. The conversion of drained and row-cropped flood-prone lands back into lands that provide ecosystem services could be a cost-effective way for taxpayers to reduce extreme flooding. In addition to receiving payments for these ecosystem services, landowners could sell carbon credits, harvest native seed, cut wetland hay for livestock, or harvest biomass for energy production. It is possible that wetland owners could see better overall returns than they ever did for row crop production on flood-prone ground.

Extreme floods occurred well before Euroamerican settlement, when the entire landscape was perennial, but not as frequently as today. Our efforts to re-perennialize the landscape can only go so far in reducing the risk of a catastrophic flood. However, climate change is likely to bring even more extreme rainfall in the spring and summer (see chapter 12). Because the increasing threats to cities, rural roads and bridges, recreational trails, and agricultural production are real, it is in everyone's economic interest to make the landscape as flood-resilient as it can possibly be.

Scientists from a wide range of disciplines have called for reduced row crop acreage and for more perennials in the landscape for environmental, economic, and human health reasons. This will require changing commodity and export policy—not just changing conservation policy. However, current farm commodity policy, which subsidizes corn and soybean production for feed, export, and biofuel, has broad political support. Conservation provisions in the Farm Bill, while helpful, have not reversed the upward trend in row crop acres. The Conservation Stewardship Program in the 2008 Farm Bill could significantly expand perennial cover on the landscape, especially if it invites broad participation, and payments are superior to those for commodities.

We now have most of the research needed to support the suggested farming and land-use changes, but two areas could use more attention. First, for row crop acres, we need a new way to measure success—water yield. What proportion of precipitation landing in a field is staying on the farm for more than, let's say, two weeks? If this could be quantified, farmers who achieve lower water yields could be compensated for this important ecosystem service. Call it another form of flood insurance.

Second, policymakers need predictive computer models that can calculate how much flood-control benefit would be achieved by replacing 10, 20, or 30 percent of the row crops in a given watershed with deep-rooted perennials. This would help determine the size and allocation of incentives to be paid for such row-crop replacement.

We also need to fully engage primary and secondary processors who control most of the grain-milling, meat-processing, and input-supply markets and who thus enjoy a high level of control over farmers' decisions and federal farm policy. Without addressing the industrial food system that drives agricultural production, the agricultural landscape cannot be re-designed (Jackson 2008). Perhaps the entrenched industrial food system will be the hardest to change.

In conclusion, the resilience we seek will be achieved bit by bit, one small stream and one micro-watershed at a time. Just as crop yields were increased year after year through incremental improvements in plant breeding, genetic engineering, and field agronomy, so can well-funded, sustained, multi-pronged, single-minded effort be dedicated to protecting our lands from future flooding.

As flood damages increase, the need for hydrological resilience grows more urgent. A re-perennialized agricultural landscape will still produce food but also will restore community values and ecosystem services that have been lost. This landscape will once again regulate and purify water, sustain soil fertility, replenish the groundwater supply, support wildlife and pollinators, and carry forward the ancient heritage of our native prairies, woodlands, and wetlands. The challenge will be to incorporate these very real ecosystem services into a market that has until now neglected and nearly destroyed them. An agricultural economy modeled on natural perennial systems will shoulder its share of the responsibility for a healthy, resilient landscape.

Note
1. Figures for "planted acres" include acres in oats, hay, corn, and beans but omit pasture and timber acreages.

References Cited
Cox, T. S., J. D. Glover, D. L. Van Tassel, C. M. Cox, and L. R. DeHaan. 2006. "Prospects for developing perennial grain crops." *BioScience* 56: 649–659.

Hawkins, A. S. 2002. "Return to Coon Valley." In *The Farm as Natural Habitat: Reconnecting Food Systems with Ecosystems*, ed. D. L. Jackson and L. L. Jackson, pp. 57–70. Washington, D.C.: Island Press.

Helmers, Matthew. 2009. Assistant Professor of Agriculture and Biosystems Engineering, Iowa State University, Ames. Personal communication, January 2009.

Iowa Corn Promotion Board/Iowa Corn Growers Association. 2009. "Iowa's ethanol/ E85 talking points" fact sheet, January 12, 2009. http://www.iowacorn.org/cms/ en/Ethanol/Ethanol.aspx.

Jackson, L. L. 2002. "Restoring prairie processes to farmlands." In *The Farm as Natural Habitat: Reconnecting Food Systems with Ecosystems*, ed. D. L. Jackson and L. L. Jackson, pp. 137–154. Washington, D.C.: Island Press.

————. 2008. "Who 'designs' the agricultural landscape?" *Landscapes Journal* 27: 23–40.

Knox, J. C. 2006. "Floodplain sedimentation in the Upper Mississippi Valley: Natural versus human accelerated." *Geomorphology* 79: 286–310.

Krug, W. R. 1996. "Simulation of temporal changes in rainfall runoff characteristics, Coon Creek Basin." *Journal of the American Water Resources Association* 32: 745–752.

Mitsch, W. J., and J. G. Gosselink. 2000. "The value of wetlands: Importance of scale and landscape setting." *Ecological Economics* 35: 25-33. Doi:10.1016/S0921-8009(00)00165-8.

National Agricultural Statistics Service (NASS). 2007. "Iowa state agriculture overview—2007." Washington, D.C.: U.S. Department of Agriculture. www.nass. usda.gov/Statistics_by_State/Ag_Overview/AgOverview_IA.pdf.

Schilling, K. E., and M. Helmers. 2008. "Effects of subsurface drainage tiles on streamflow in Iowa agricultural watersheds: Exploratory hydrograph analysis." *Hydrological Processes* 22: 4497–4653. DOI: 10.1002/hyp.7052.

Smith, Daryl. 2008. Professor of Biology and Director, Tallgrass Prairie Center, University of Northern Iowa, Cedar Falls. Personal communication, June 2008.

Tilman, D., J. Hill, and C. Lehman. 2006. "Carbon-negative biofuels from low-input high-diversity grassland biomass." *Science* 314: 1598–1600.

Gerald E. Galloway

25 The Great Flood of 1993
Did We Learn Any Lessons?

IN JULY 2008, following the Midwest floods, the United States Senate Environment and Public Works Committee held a hearing to review what had happened and what might be improved in management of flood risk (U.S. Senate 2008). The chairman of the committee, Senator Barbara Boxer, questioned whether any lessons had been learned from the floods of 1993 and asked the Hon. John Paul Woodley, assistant secretary of the U.S. Army, if he was familiar with a report that had been prepared for the White House following those floods (IFMRC 1994). Mr. Woodley indicated he was aware of the report, that he had put some of its recommendations to use, and that those recommendations that had been carried out brought positive results in 2008. Senator Boxer asked him to once again review this report to determine what recommendations had not been addressed. This chapter looks back at the report that came out of the 1993 floods and considers what has happened since it was presented to the President of the United States in 1994.

From the fall of 1992 into the spring of 1993, the Midwest experienced greater than normal precipitation. In Iowa, the November–April rainfall was the second greatest in the 121 years of record (IFMRC 1994). So, when the summer of 1993 arrived, conditions were ripe for the creation of a major flood event—soils were saturated, and many streams and rivers were already above their normal levels for early summer. Then, an unusual weather pattern sat over the region from June through late July and brought rainfall amounts that exceeded all historical records (IFMRC 1994). Small rivers began to overflow onto nearby lands and send their increasing flows downstream to fill larger rivers. By the middle of July, the nation was witnessing the devastating effects of the Great Flood of 1993 on eight midwestern states (figure 25-1).

As floodwaters began to recede, total damages and recovery costs were estimated to exceed $22 billion, the largest flood loss in the United States in the 20th century. Thirty-eight deaths were attributed to the flooding. Over 100,000 homes were damaged. Crop losses from flooding and saturation of the soil reached $2.5 billion. Factories were closed and businesses were disrupted as transcontinental highways and railroads were severed by floodwaters and workers left the area. Navigation on the Mississippi River was halted, and loaded barges were tied up along the banks. Already-high levels of family stress grew higher. Six months after the floods began, water was still being pumped from fields in low-lying areas, and thousands of acres of farmland remained covered with sand and would never be farmed again (IFMRC 1994).

FIGURE 25-1 The Des Moines River in flood, July 1993. *Photograph by Andrea Booher, FEMA.*

The floods of 1993 brought near-continuous television coverage to the region. As people across the country sat down to their evening meal, night after night they saw pictures of more inundated land, the agony of families leaving their flooded homes, and possessions under muddy waters. Later, they watched victims return to their destroyed homes, communities, and farmlands. Such media coverage raised concerns about flooding across the nation. It spurred President Bill Clinton and members of the Congress from the Midwest to ask, "How did this happen, and what can be done to prevent it from happening again?"

In 2008, much of the Midwest was again under water and struggled to understand how an event so similar to the 1993 floods, with its multibillion-dollar devastation, could once again occur. To answer this question, let's back up to consider the federal government's response to the 1993 floods.

In November 1993, the White House established the Interagency Floodplain Management Review Committee (IFMRC) within the Executive Office of the President. The committee was charged with finding out why the floods had occurred and what should be done to prevent flood losses in the future. Six months later, the committee reported (figure 25-2) to the president that:

- The floods of 1993 were the result of a significant rainfall event. Floods of that magnitude occurred in the past, and floods like the 1993 event would continue to occur.
- Locally constructed levees failed at an alarming rate, were poorly sited, and would fail again. States had provided little oversight of the construction and operation of such structures. In contrast, federal flood projects worked essentially as designed and significantly reduced the damages to population centers, agriculture, and industry. However, the combination of federal and local levees and other structures represented an uncoordinated aggregation that did not ensure reduction in the vulnerability of those living and working in the floodplain. Even with levee protection, people and property in the floodplain continued to face the residual risk of potential overtopping or failure of levees.
- Flood damage-reduction projects, navigation works, and land use practices throughout the basin caused significant loss of habitat and ecosystem diversity and adversely altered the remaining bottomland habitat. This loss of wetlands and upland cover and other landscape modification throughout the basin significantly increased runoff. Conversely, upland watershed treatment and restoration of upland and bottomland wetlands could reduce flood stages.

- The level of protection provided to many existing population centers and critical to infrastructure such as hospitals, water treatment facilities, and fire stations was inadequate.
- People living in the floodplain and eligible for flood insurance were not buying it. Fewer than 30 percent of those required to purchase insurance under the National Flood Insurance Program (NFIP) had policies (IFMRC 1994). [The Federal Emergency Management Agency (FEMA) recently indicated that fewer than 10 percent of those affected by the 2008 Midwest flood carried flood insurance (Maurstad 2008).]

In response to its conclusions, the committee offered more than 50 recommendations, beginning with its call to avoid future development of a floodplain where alternatives exist. It also proposed that every effort should be made to reduce vulnerability of existing development or necessary new development on floodplains. In order of priority this should be done through permanent evacuation of highly flood-prone areas, especially those with a history of flooding; establishment of flood warning systems to allow for relocation of people and valuable items before floods arrive; flood proofing of structures remaining in the floodplain; creation of additional natural and artificial storage to hold upstream water where it falls; and, when environmentally, socially, and economically justified, construction of new, adequately sized and maintained levees and other flood damage-reduction structures.

The committee also indicated that the president should propose a Floodplain Management Act to delineate federal, state, tribal, and local responsibilities for floodplain management; provide fiscal support for state and local floodplain management activities; establish states and tribes as the nation's principal floodplain managers; and charge states and tribes with ensuring the proper siting, construction, and maintenance of non-federal levees.

Noting the problems with the acceptance and purchase of flood insurance, the committee recommended that FEMA take vigorous steps to improve the marketing of flood insurance, enforce lender compliance rules, seek state support of insurance marketing, reduce the amount of post-disaster support to those who were eligible to buy insurance but did not, and require those who live behind levees to purchase actuarially based insurance to cover the residual risk they face.

Given the broad-ranging recommendations of the 1994 report, one might wonder what happened to the recommendations and knowledge generated by the report. How, in the wake of the report, could the 2008 floods have brought even more destruction in parts of the upper Mississippi Basin than those of 1993?

During and immediately following the 1993 floods, the national focus was on taking care of those harmed by the floods and on rebuilding the region's economic viability. Little emphasis was placed on long-range changes directed toward reducing future extreme floods.

Recommendations on dealing with the future were the province of the IFMRC. When the committee's report was published in June 1994, it received

FIGURE 25-2 Cover of the major governmental report addressing the 1993 midwestern floods (IFMRC 1994). The committee producing this report was chaired by the author. Few of the recommendations in this science-based report were adopted. The cover photograph shows the flooding Missouri River in 1993. *Document cover photograph by the Missouri Department of Conservation.*

headlines across the Midwest. There were editorials and speeches in support of doing something about the "flood problem." Congress took up and acted on some elements of insurance reform. It also began hearings on other committee recommendations. However, the November 1994 congressional election shifted control of the Congress and, as a result, control of congressional committees changed hands. When the new Congress took office in 1995, the attention of the House and to a lesser degree the Senate shifted to carrying out new agendas. By then the memory of the 1993 floods had begun to fade in the minds of members of Congress and, judging by the lack of congressional committee action, interest in doing something about flooding almost disappeared. The governors and legislators of the eight affected midwestern states also found other topics to worry about, hoping that the impetus for major flood-related efforts would come from the federal government.

In the White House, President Clinton's staff continued to work on carrying out many of the committee's recommendations, but turf squabbles among agencies and emphasis on other priorities led to the demise of any real flood-mitigation actions.

In 2005, when Hurricane Katrina struck the Gulf Coast and devastated New Orleans and several Mississippi communities, interest in flooding returned, and Congress again held hearings on what might be done. The 1994 report was dusted off and re-presented to the House. However, the Bush administration's focus was on recovery rather than on developing long-term approaches to dealing with flood problems. Without direction from the White House, federal agencies went their own ways, devoting most of their efforts to addressing New Orleans' recovery rather than long-term national issues. Katrina did point out the fragility of levees and related flood-control structures and caused both FEMA and the U.S. Army Corps of Engineers to examine more closely the status of levees across the nation. Surprisingly, the scrutiny of local levees was not welcomed and instead brought protests from many local agencies responsible for levees.

Then came the 2008 Midwest floods, with their harsh reminder that significant floods continue to occur and that steps need to be taken to lessen the damages they bring to people and communities. Immediately following the 1993 floods, FEMA, in cooperation with state agencies, initiated a significant effort to relocate or remove floodplain homes that had been repetitively damaged along with homes that would obviously be flooded in coming years. This approach represented a marked change from the past. It led to the movement of sections of small communities and, in one case, an entire town, Valmeyer,

Illinois. But, for the most part, over the 15 years between floods, governments failed to take action to lessen the probability or destructive capability of such an occurrence. Some communities like Des Moines, in the absence of federal action, took it on themselves to address their problems, but many public officials either ignored the threat of future major floods or were ignorant of their potential severity. That ignorance was echoed by flood victims who did not understand that they were at risk and that the levees protecting them had not been designed to handle large floods. The words "nobody told me" were repeated time and again and resound in flooded communities even today. Judging by insurance participation statistics, few even now see it as their responsibility to protect themselves against the possibility of floods.

At a time when the nation faces grave fiscal crises, infrastructure in need of repair, climate change that may bring increased flooding, and a continuing over-reliance on structural solutions that leave residual risk, we can no longer afford to ignore the lessons of major floods. In 1994, the IFMRC proposed widely accepted, solid, science-based proactive measures to reduce, across the nation, future flood losses and the trauma they bring. Yet, because memories of floods fade rapidly and dealing with many of these recommendations would have brought political push-back, few of the measures were adopted. Now, with both Katrina and the Midwest floods of 2008 still on our minds, will we once again avoid the issue? Will we again return to business as usual and fail to remember the harsh lessons of our 21st-century floods? Will we again leave future generations to deal with the challenges that we should have met? We owe it to them to do better.

References Cited

Interagency Floodplain Management Review Committee (IFMRC). 1994. *Sharing the Challenge: Floodplain Management into the 21st Century.* Report of the Interagency Floodplain Management Review Committee to the Administration Floodplain Management Task Force. Washington, D.C.: U.S. Government Printing Office.

Maurstad, D., Federal Emergency Management Agency (FEMA). 2008. Presentation to National Association of Flood and Stormwater Management Agencies, August 27, 2008. Napa, California.

U.S. Senate Committee on Environment and Public Works. 2008. "The Midwest floods: What happened and what might be improved for managing risk and responses in the future." July 23, 2008. http://epw.senate.gov/public/index.cfm?FuseAction=Hearings.Hearing&Hearing_ID=2e21708d-802a-23ad-45ed-f9084320af55.

Epilogue

~~~~~~

My OFFICE WINDOW overlooks the Iowa River in downtown Iowa City. Sometimes I glance up from work to see flocks of gulls the size of small clouds swooping over the water, or dozens of feeding swallows, or soaring bald eagles searching for fish. Once, even here in the city, an otter bobbed in the water below my window. Often people fish along the river's gravel shores, chatting and hauling in buckets of catch. But I have also heard the shouts of rescue workers probing the river bottom for the bodies of boaters who were swept over the nearby dam, and I have watched Norway rats the size of cats scurry along the river's banks at dusk. I have seen what this river can give and what it can take. I have heard the river's divergent voices.

The 21st century has been dubbed the water century, a time when fresh water will take the place of oil as the most highly sought-after resource. Many predict that the search for adequate fresh water for Earth's growing billions of people will shape foreign policy and cause wars, as well as human suffering. Already dozens of nations lack the water they need. This may not be a comfort to people recovering from flooding, but consider this: sufficient water and rich topsoil are Iowa's and the Corn Belt's most valuable natural resources, the envy of nations around the world. The way we exploit their benefits has diminished these resources, twisting them into interacting environmental problems, our water gushing away with our rich topsoil, the topsoil polluting our waterways. But these two commodities remain elementary assets. Our task is to find ways to ensure that both serve as healthy, balanced assets that nurture, rather than harm, life.

How might this be done? First, we need to accept that floods will continue, we have not yet seen the biggest floods, floods and their damages are becoming larger, and climate change may further intensify these trends. There is no silver bullet for preventing or totally controlling floods or eliminating their destruction. But we can learn to live with floods, trusting that their size and destructive power can indeed be moderated.

Living with floods involves two broad activities: better managing the risks and taking steps to reduce our vulnerability, and better managing the landscape to reduce the magnitude and destructive power of floods. The first incorporates practical actions such as flood-proofing valuable structures or moving them out of the floodplain. Section IV described a number of other vulnerability-reducing activities, including revamping of floodplain regulations, stretching these to consider the entire watershed, following through with long-term flood mitigation strategies, and implementing careful and well-planned engineered constructs such as flood barriers. Ongoing research is sure to present us with greater knowledge (see plate 11) and more options for reducing vulnerability.

The second activity, managing the landscape to reduce flood magnitude, requires restoring our land so that it will absorb and hold more water, as it did in previous centuries. Specific techniques for doing so are increasingly being incorporated in urban areas. We should not ignore the power and importance of individual and community actions. As suggested in chapter 23, each one of us can take steps toward healing our landscape and restoring its flood resilience. Just as a river's excess flood flows are constructed drop by drop, so each of us can lessen these flows drop by drop—through the way we manage our own land.

But we also need to consider the vast stretches of annual row crops that now cover about two-thirds of our state and shed water as well as soil and chemical pollutants. Their vast coverage and intensive manipulation along with the elimination of perennial vegetation have created a fragility in our landscape, one matched elsewhere around the globe by ongoing loss of biodiversity, increasing environmental contaminants, and other pressures. For the sake of today's citizens as well as those to come, the Corn Belt's landscape must once again become a provider of environmental health, rather than be narrowly viewed as a site only of agricultural commodity production. The concept of ecological services lands is now being implemented in hundreds of projects around the world, with China investing $100 billion to create ecological function zones that support flood control throughout the country (Ellison 2009). Lands throughout the Corn Belt have the potential to grow both "ecological services" and food in quantity. If we adopt this approach, the benefits and products provided by our landscape would be multiplied, with food and crop production being matched by the land's safeguarding of our state's environment and citizens' health. Linking the two would shore up our landscape's ability to function sustainably and moderate floods for years to come.

Reconsidering current land use patterns may be the last thing that people and communities want to do when they are recovering from destructive flooding. At such times, human nature yearns for the comforting and familiar and seeks to return to the way things once were. But returning to business as usual is not listening to the voices of nature. Extreme floods that play havoc with human lives and the land are telling us that the pre-flood status quo is not working, that human relationships with the land are other than they should be. Destructive floods need to be seen as a symptom of some larger imbalance rather than as isolated events.

Addressing the larger causes of flooding would require a major rethinking of our relationship with the landscape, one that would affect all aspects of our society and may well entail large-scale policy changes. We would need to surmount institutional and policy inertia, as well as challenge the power of special interests and lobbies. These are no small tasks. But consider the alternative. What have extreme floods cost us so far? What will they cost in the future? Each time we recover from a weather disaster such as a major flood or tornado, we invest tremendous resources and funds in reactively "returning to normal"—resources and funds that could have been used proactively to prevent future destruction. These costs may well be magnified (and our ability to act proactively reduced) by the repercussions of future climate change. At some point, our reactive investments of time, money, and community spirit may be so large that proactive efforts to reduce flood damages will no longer be possible. If we reach that point, we will have committed ourselves to ongoing cycles of destruction and response. The door of opportunity is now open. The wise choice is to walk through and act now, while we can, rather than wait for a future when our options may be fewer.

If we open ourselves to embracing new flood-reducing approaches and solutions, we can address multiple environmental problems and reap multiple environmental benefits at the same time. By seeking to restore our landscape, establish more water-holding perennial cover, reconsider floodplain land uses, and adopt new watershed-based floodplain management approaches, we will not only decrease the size of the floods that surely will come. We will also improve the quality of our water, create habitat for diverse native plants and animals, and increase outdoor recreational opportunities that can draw us into healthier lifestyles. These benefits in turn will make our region a more attractive place to live, one that holds onto our children and attracts new workers and economic investment. At the same time, we will be doing our part to improve our planet's health so it can better meet future environmental challenges.

Iowa's 2008 floods were not caused by human activity; with the tremendous quantities of water gushing down our waterways, floods were sure to occur. But they were giving us a message about the sustainability of our landscape, just as the 1993 floods did. Will we listen to the message? The Iowa and Cedar River watersheds are the subject of abundant flood research; might they also become demonstration sites for new approaches to flood mitigation, blending economic and environmental health in ways that later could be emulated throughout the region? Could Iowa—in the heart of the Corn Belt—become a model for lessening water's negative power by tying agricultural practices to their water consequences? The next time floods strike, might we look with pride at how a healthy and resilient landscape, wise planning, and well-thought-out mitigation strategies reduced the flood damages?

If we act wisely, the 2008 floods will be viewed not as an endpoint or time of loss but rather as a new beginning stimulated by challenges and possibilities. We have things to do. Let's get busy.

Reference Cited

Ellison, K. 2009. "Ecosystem services—out of the wilderness?" *Frontiers in Ecology and the Environment* 7: 60.

Notes on Contributors

Lynn M. Alex (chapter 13) is director of education and outreach (and formerly research archaeologist) at the University of Iowa Office of the State Archaeologist, where she has been since 1991. She presents archaeological research to the public through writings, presentations, and development of school curricula. The proliferation of research in this field following Iowa's 1993 floods heightened her awareness of the impact of floods on people in ancient history and their remains. Her book *Iowa's Archaeological Past* (2000) is the definitive guide to Iowa archaeology.

Joe Alan Artz (chapter 13) joined the University of Iowa Office of the State Archaeologist in 1989, where he is the director of the Geospatial Program. He divides his time between computer mapping and modeling the distribution of archaeological sites on the landscape, and studying stream valleys to find buried archaeological sites and to reconstruct floodplain landscape evolution through time. These studies inevitably incorporate flooding. His fieldwork has taken him to sites throughout the Great Plains and Midwest and to other countries.

A. Allen Bradley, Jr. (chapters 1 and 6), is an associate professor of civil and environmental engineering and research scientist at IIHR-Hydroscience & Engineering, the University of Iowa, where he has been since 1994. He combines a broad understanding of hydrology with research on hydrologic extremes (both floods and droughts). For the past decade, Bradley has been working with the National

Weather Service to assess and improve long-range river forecasts of flood and drought conditions, efforts that have made him one of the nation's experts in hydrologic forecast verification.

Michael Burkart (chapter 8) has been examining the effects of agricultural systems on hydrology in the Mississippi drainage basin for the last 15 years, first as research leader of the Agricultural Land Management Unit at the National Soil Tilth Laboratory in Ames, Iowa, and since his retirement as a consultant to the Des Moines Water Works, where he works on the relationship between landscape modifications, river discharge, and water chemistry in the Raccoon River basin. He previously studied agriculture's effect on water chemistry for the U.S. Geological Survey.

John Castle (chapter 10) has worked for 32 years at Coralville Lake for the Rock Island District of the U.S. Army Corps of Engineers, where since 1991 he's served as operations project manager. Castle is responsible for all activities on the Corps' Coralville property, including operation of the dam and reservoir, a responsibility he held during the 1993 and 2008 floods. In carrying out his tasks, Castle interacts regularly with the Corps' Water Control Section in Rock Island.

Richard Cruse (chapter 16) is a professor of agronomy and director of the Iowa Water Center at Iowa State University. As team leader for the Iowa Daily Erosion Project (now on the Web as the Iowa Mesonet Network), he built a team with access to the tools needed to address important questions related to spatial rainfall and runoff patterns and their influence on water's flow, erosion, and soil water storage. Throughout his career, Cruse has researched soil management issues (e.g., tillage and crop systems) and their relationship to soil and water resources.

David Eash (chapter 7) is a hydrologist at the U.S. Geological Survey's Iowa Water Science Center in Iowa City, where he has been employed since 1986. He works with streamflow statistics, focusing in particular on floods and droughts. His primary job is to perform flood-frequency estimation for the entire state—including both gaged and ungaged sites. Eash was the primary hydrologist performing flood-frequency estimation for both the 1993 and 2008 floods in Iowa. He also does flood-profile delineations to document water-surface elevations along river reaches for significant floods in the state.

Barbara Eckstein (chapter 4) came to the University of Iowa in 1990 and is a professor of English and associate provost for Academic Administration. In the latter position, she represents the university's academic mission in space concerns across campus. She was a member of the university's 2008 flood response committee and continues to work on flood recovery efforts. She is broadly interested in sustainability issues and the quality of the Iowa River, and in 2007–2008 she developed an educational program for the lay public that incorporated guest speakers and field tours to explore this theme.

Engineer **Richard A. Fosse** (chapter 3) is public works director for the City of Iowa City. He was incident commander for the city during much of the flood. As such, he was responsible for facilitating the city's day-to-day responses to rising floodwaters and, later, the massive cleanup. He played a similar role during the 1993 flood. He was well prepared for this task by both his 25 years of service with Iowa City's municipal government and his prior work in floodplain management with the Iowa Department of Natural Resources.

Gerald E. Galloway (chapter 25), for the past 40 years, has focused on water resources policy and on floods and their prevention. He is now a professor of engineering and affiliate professor of public policy at the University of Maryland. He also consults on water resources around the world. Prior to these efforts, he served in the U.S. Army Corps of Engineers, retiring after 38 years as a brigadier general. In 1993–1994, he led the White House Study of the 1993 Mississippi River floods, producing what is now called the Galloway Report.

Keri Hornbuckle (chapter 17) is a professor and the chair of civil and environmental engineering and a professor of occupational and environmental health at the University of Iowa. She grew up in Cedar Rapids, where she and her students started sampling flood-related pollutant residues in soils in August 2008. She has been studying the transport and fate of toxic chemicals in the environment for over 20 years, in particular fragrances, PCBs, and pesticides in the Great Lakes and other surface waters of North America.

Laura Jackson (chapter 24) is a professor of biology at the University of Northern Iowa, where she has worked since 1993. She teaches ecology and conservation biology and coordinates a graduate program in ecosystem management. She has been studying the restoration of biological diversity in agricultural landscapes for many years. Her broad interest in native and agricultural grasslands and their ecological functions stems from growing up on the Kansas prairies. She is coeditor (with Dana Jackson) of *The Farm as Natural Habitat: Reconnecting Food Systems and Ecosystems* (2002).

Douglas M. Johnston (chapter 21) is a professor and the chair of the Departments of Community and Regional Planning and Landscape Architecture at Iowa State University (ISU). He joined the ISU faculty in 2007 after teaching for more than 20 years at the University of Illinois. Teaching and research throughout his career have assumed a systems perspective of land use and environmental problems, particularly water resources and land management. His work examines changes in ecological and economic value of land from flood management, particularly the impact of local urban best-management practices for reducing stormwater flow in the larger watershed.

Dennis Keeney (chapter 24) is a senior fellow at the Institute for Agriculture and Trade Policy in Minneapolis, a position he assumed in 2000 after retiring from directing the Leopold Center for Sustainable Agriculture at Iowa State University. He has worked in landscape and sustainability issues, in particular agricultural policy and impacts, throughout his career, with an ongoing interest in watersheds and flooding. Keeney also brings a personal interest to his chapter: 2008 floodwaters permanently damaged 150 acres of his family farm by covering them with up to a foot of sand and gravel.

Dana Kolpin (chapter 17) is a research hydrologist at the U.S. Geological Survey's Iowa Water Science Center in Iowa City. Since 1986, he has conducted numerous water quality investigations across the United States. His recent efforts have addressed pharmaceuticals and other under-investigated chemical contaminants. His research has included several flood-related studies including wells affected by the 1993 flood, the 1998 flooding of the Nishnabotna River in western Iowa, and chemical transport during the 2008 floods. He emphasizes that the transport of chemicals in floodwaters is an important aspect of extreme hydrologic events.

Witold F. Krajewski (chapter 2) is a professor of civil and environmental engineering and a research engineer at IIHR-Hydroscience & Engineering, the University of Iowa. He came to the university in 1987 from the National Weather Service. His hydrological research focuses on remote sensing of rainfall using radar and satellites. A highly respected international expert in rainfall research, he has developed field observatories and methodologies for assessing uncertainty in rainfall estimates, with the goal of improving high-resolution monitoring and forecasting of rainfall and flooding.

John M. Laflen (chapter 16) was an agricultural engineer with the U.S. Department of Agriculture, Agricultural Research Service, in multiple locations from 1960 until he retired from the National Soil Erosion Laboratory, Lafayette, Indiana, in 2000. His work focused on prediction and control of soil erosion. He contributed to some of the earliest and largest computer models on this subject. His erosion focus ties him directly to flooding, which accounts for the largest erosion events. He has worked with coauthor Richard Cruse for more than 25 years, and he served on coauthor Hillary Olson's M.S. committee.

Linda Langston (chapter 5) is a member of the Linn County Board of Supervisors. She was chair in 2008 and as such led the supervisors in their oversight of Linn County's response to the flooding. She attributes the success of this multifaceted oversight to the positive relationships that existed among the multiple participants at the Emergency Operations Center. But equally important in successfully "making 1,000 decisions a day" were her leadership training and prior experience as a museum director, business owner, and participant in the Harvard-Kennedy School of Government.

Architect **Rodney Lehnertz** (chapter 4) is director of planning, design, and construction in Facilities Management at the University of Iowa, where he has been since 1994. His responsibilities include campus master planning, capital project management, space management, and leadership of the Campus Planning Committee. During and following the 2008 floods, he participated in daily meetings of the university's flood response committee and was responsible for coordinating volunteers and working with the media. His coauthored book, *The University of Iowa Guide to Campus Architecture* (2006), has proven helpful in recovery planning for various university buildings.

Ricardo Mantilla (chapter 2) came to the University of Iowa in 2008 as a post-doctoral associate at IIHR-Hydroscience & Engineering, hoping to test the flood-causation mechanism he had developed at the University of Colorado for his doctoral dissertation. A few months later, the 2008 floods allowed him to do so. His expertise in water's movement through a landscape, with timed convergences causing or enhancing flooding, complements his coauthor's expertise in the space/time distribution of rainfall, allowing the two to promote a new theory of flooding as explained in their chapter.

Cornelia F. Mutel (book editor, author of Introductions, chapter 14, and Epilogue) is on staff at IIHR-Hydroscience & Engineering, the University of Iowa, where she tends archives and since 1990 has written on the history of her home institute and on global change issues. An ecologist by training, she focuses on writings that interpret science and environmental issues for lay audiences. She is the author or editor of a dozen books, including several on landscape, natural history, and restoration ecology, most recently *The Emerald Horizon: The History of Nature in Iowa* (2008), Iowa's first comprehensive natural history.

A. Jacob Odgaard (chapter 22) is a professor of civil and environmental engineering and research engineer at IIHR-Hydroscience & Engineering, University of Iowa, where he's been since 1977. His expertise in river mechanics in general, and river channel stabilization and erosion more specifically, has led him into research and consulting efforts in improving the flood conveyance of river channels, in particular sediment management. His observations of hundreds of rivers around the world help him predict the natural behavior of rivers and work with their natural tendencies in order to minimize flood destruction.

Hillary Olson (chapter 16) is program coordinator for the Iowa Water Center at Iowa State University, where she is in charge of outreach and education for the center's water- and soil-conservation issues. Prior to coming to Ames in 2007, she was a soil conservationist for the U.S. Natural Resources Conservation Service and completed graduate training in soil science. She remains an active leader of the Soil and Water Conservation Society, grew up and still lives on a working farm, and has abundant practical experience with soil problems and processes.

Daniel Otto (chapter 15) is a professor of economics at Iowa State University (ISU). Since arriving at ISU in 1981, he has conducted extension programming on rural economic development, taught public finance, and focused broadly on rural development and regional economics. He regularly performs economic impact assessments for Iowa's state and local governments, addressing diverse issues such as the economic impact of tax policies and weather events like floods and droughts. He worked with state officials and others to perform the economic assessment of Iowa's 1993 floods and now is doing the same for the 2008 floods.

John Pearson (chapter 19) is a botanist and ecologist for the Natural Areas Inventory program of the Iowa Department of Natural Resources. Since coming to Iowa in 1985, he has worked with plant community classification, inventory of prairies and fens, threatened and endangered species, state nature and cultural preserves, ecological management of sensitive habitats on state lands, and conflicts between development and environmental preservation. His wide-ranging interests—including ecology, geology, archaeology, lichens, and insects—and his extensive field experience and many interpretive presentations make him one of Iowa's most recognized naturalists.

Wayne Petersen (chapters 9 and 23) has 33 years of field experience with soil and water conservation in Iowa. He helped establish urban water conservation as an area of concern and action, teaching Iowans new ways of handling precipitation to control both water quality and runoff. He is now urban conservationist for the Iowa Department of Agriculture and Land Stewardship, where he coordinates the statewide Urban Conservation Program and provides technical assistance to Iowa's Soil and Water Conservation Districts. Earlier, he played a similar role with the U.S. Natural Resources Conservation Service.

Jack Riessen (chapter 20) retired in 2009 from the Iowa Department of Natural Resources, where he worked in a variety of water-related issues including hydrology, hydraulics, drainage, and water quality. At various times throughout his career he oversaw the state's floodplain management program, coordinated the National Flood Insurance Program in Iowa, and headed the Water Quality Bureau. Most recently, he served as senior technical expert on water issues in the director's office and worked on the state water plan, including leading an expert panel on floodplain management issues.

Robert F. Sayre (chapter 11) is an emeritus professor of English and taught at the University of Iowa for 33 years before retiring in 1998. Throughout his career, he has been intensely interested in landscape history and human interactions with nature. He taught courses on landscape and literature and edited three books on Iowa's prairies and landscapes. Sayre is a lifelong sailor and lover of oceans, lakes, and rivers. He and his wife, Hutha, have canoed many parts of the Iowa, Cedar, and other Iowa rivers, and he continues to study how human activity has transformed natural landscapes.

Eugene S. Takle (chapter 12) is a professor of agricultural meteorology and a professor of atmospheric science at Iowa State University (ISU). At ISU, where he's been since 1971, he also directs the multidisciplinary Climate Science Initiative. He is a leader in regional climate change research (especially applied to the Midwest) and currently codirects ISU's Regional Climate Modeling Laboratory, which is a national leader in regional climate modeling and development of future regional climate scenarios for North America.

Peter S. Thorne (chapter 18) is a professor and director of the Environmental Health Sciences Research Center at the University of Iowa. His major research focus on bioaerosols and human health has always included mold, but his mold research intensified following the 1993 floods and Hurricane Katrina (2005), at which times he assessed exposure to and human health effects of molds perpetuated by the floods. The 2008 floods allowed him to do more of the same but also to relate his research work directly to public health issues and education, an applied outreach area that he greatly enjoys.

Nathan C. Young (chapter 22) is an associate research engineer at IIHR-Hydroscience & Engineering and an adjunct associate professor of civil and environmental engineering, University of Iowa. He joined the university in 2008 and his training, research, and experience in river mechanics and floodplain inundation mapping have taught him to understand the flows of rivers during normal and flood conditions and their reshaping by floodplain structures. In 2008, he started to use computerized numerical models to reconstruct that year's flood flows from the Coralville Dam through Iowa City, an effort that will support flood mitigation strategies.

Index

~~~~~~

Rapids; Coralville; Coralville dam and reservoir; hydrographs; Iowa City; Marengo; nature; precipitation; runoff; University of Iowa

Galloway report, 227, 229–232

health hazards: airborne, 163–170
hydrographs, 4, 13, 15, 21–23, 57, plate 14
hydrologic cycle, 26, 87–89, 215–217

Interagency Floodplain Management Review Committee (IFMRC). *See* Galloway report
Iowa City, Iowa, xiii–xv, xix; 100-year and 500-year designations, 63, 65–66, 187–188; and 1993 flood, 13–15, 37, 104, 188; and 2008 flood events, 31–37, 55–58, 98, 100–101, 158, plates 10 and 11; and 2008 flood discharges and stages, 13–15, 19, 22, 57, 63–64, 67, 75, 100, 134, 187–188; floodplain development and management, 39, 104, 106–109, 187–188; historic floods, xi–xii, 60, 63–64, 67, 103. *See also* Coralville dam and reservoir; floodplain; Iowa River; Johnson County; University of Iowa
Iowa River: and 2008 flood, 22, 26–29, 55, 63–64; watershed, xiii–xv, 27, plates 1 and 12. *See also* Coralville; Coralville dam and reservoir; floods of 1993; floods of 2008; Iowa City; Marengo; streamgage stations

Johnson County, Iowa, xiii–xiv; and 2008 flood, 32–33, 136, 143–144, 150, plate 9; archaeological sites, 125–128; vegetation, 174, 176–177, plates 7 and 8. *See also* Coralville; Coralville dam and reservoir; floods of 2008; Iowa City; University of Iowa

levees. *See* flood control, barriers
Linn County, Iowa, xiii–xiv; and 2008 flood, 45–50, 134, 143–144, 150, plate 9; archaeological sites, 125–126; vegetation, 176–177, plates 7 and 8. *See also* Cedar Rapids; Cedar River; floods of 2008

Marengo, Iowa, xiii–xiv, 19–20, 23, 36, 63–64, 76, plate 14
mold. *See* health hazards

National Flood Insurance Program (NFIP). *See* floods, insurance
nature and flood effects, 171–177, 202, 229

perennial cover and flood reduction, 80–84, 152–153, 217–224. *See also* agriculture; Conservation Reserve Program; prairies; wetlands
Polk County, Iowa, 92–93. *See also* Raccoon River basin
prairies, 174, 176–177, plate 7; hydrology, 74, 77–82, 87–89, 217–218, 223. *See also* soils; wetlands
precipitation, 9, 88–90, 206, 211–212; in 1993, 11–12; in 2008, 11–12, 20, 23–29, 55–58, 112–113, 150, plates 2–4, 13, and 14; predicted future, 113–115; recent trends of, 112, 115

Raccoon River basin, Iowa, 79, 81–84
rain gardens. *See* bio-retention structures
recurrence-interval estimation. *See* floods, frequency estimation
reservoirs, 29, 104, 186. *See also* Coralville dam and reservoir
river, defined, xvi
river discharge: defined, 4, 53; modern increase in, 78–84, 220; and time-compounded flows, 26–29. *See also*

Selected Bur Oak Books